Springer Tracts in Mechanical Engineering

Springer Tracts in Mechanical Engineering (STME) publishes the latest developments in Mechanical Engineering—quickly, informally and with high quality. The intent is to cover all the main branches of mechanical engineering, both theoretical and applied, including:

- Engineering Design
- Machinery and Machine Elements
- Mechanical structures and stress analysis
- Automotive Engineering
- Engine Technology
- Aerospace Technology and Astronautics
- Nanotechnology and Microengineering
- Control, Robotics, Mechatronics
- MEMS
- Theoretical and Applied Mechanics
- Dynamical Systems, Control
- Fluids mechanics
- Engineering Thermodynamics, Heat and Mass Transfer
- Manufacturing
- Precision engineering, Instrumentation, Measurement
- Materials Engineering
- Tribology and surface technology

Within the scopes of the series are monographs, professional books or graduate textbooks, edited volumes as well as outstanding PhD theses and books purposely devoted to support education in mechanical engineering at graduate and post-graduate levels.

Indexed by SCOPUS and Springerlink.

To submit a proposal or request further information, please contact: Dr. Leontina Di Cecco Leontina.dicecco@springer.com or Li Shen Li.shen@springer.com.

Please check our Lecture Notes in Mechanical Engineering at http://www.springer.com/series/11236 if you are interested in conference proceedings. To submit a proposal, please contact Leontina.dicecco@springer.com and Li.shen@springer.com.

More information about this series at http://www.springer.com/series/11693

Fei Wang · Zhenping Weng
Lin He

Comprehensive Investigation on Active-Passive Hybrid Isolation and Tunable Dynamic Vibration Absorption

Springer

Fei Wang
China Ship Scientific Research Center
Wuxi, Jiangsu, China

Lin He
Institute of Vibration and Noise Research
Wuhan, Hubei, China

Zhenping Weng
Wuhan Second Ship Design
and Research Institute
Wuhan, Hubei, China

ISSN 2195-9862 ISSN 2195-9870 (electronic)
Springer Tracts in Mechanical Engineering
ISBN 978-981-13-3055-1 ISBN 978-981-13-3056-8 (eBook)
https://doi.org/10.1007/978-981-13-3056-8

Library of Congress Control Number: 2018959256

This Springer imprint is published by the registered company Springer Nature Singapore Pte Ltd.
The registered company address is: 152 Beach Road, #21-01/04 Gateway East, Singapore 189721, Singapore

Acknowledgements

The authors would like to sincerely thank Prof. Yu Mengsa for the precious advices during the review of this manuscript and also appreciate the efforts Dr. Yin Zhiyong has made in designing adaptive dynamic vibration absorber.

Contents

List of Figures

List of Tables

Abstract

Mechanical noise is the main source of noise in warships at low speeds. Traditional passive control technology cannot provide effective low-frequency vibration control while ensuring that the volume and weight meet engineering requirements. Active control or semi-active control can effectively control low frequency vibrations through vibration cancelation or structural adaptive changes.

This book aims at controlling low-frequency vibration transmission of typical power equipment and piping system of ships through studying the active and passive hybrid vibration isolation and adjustable frequency dynamic vibration absorption technology. The main research contents and innovations of this book are as follows:

(1) The active and passive hybrid vibration isolator based on Halbach magnetic array has the characteristics of small size, light weight, and low power consumption for the same operating force requirements. Tested per unit of current can produce 300 N power, consistent with the theoretical calculation result. The single-frequency excitation test shows that the vibration isolation effect of active and passive hybrid vibration isolation is 11 dB higher than passive vibration isolation. The digital power amplifier designed and developed solves the problems of large, heavy, and severe heat dissipation of analog power amplifiers.
(2) According to the vibration characteristics of WD618 diesel engine, an adaptive feed-forward control strategy is proposed and a distributed feed-forward control hardware system is studied. The diesel engine vibration isolation test under practical application environment verified that the passive vibration isolation reduces the base vibration by 10 dB in the frequency range of 1000 Hz. The active–passive hybrid vibration isolation can further reduce the base vibration in the frequency band below 70 Hz by 4 dB, and the deck vibration below 100 Hz further drops 7 dB.
(3) The three-way dynamic vibration absorption theory is used to design the piping system adjustable frequency dynamic shock absorber, which has the advantages of compact structure, convenient disassembly and assembly, low power

consumption, wide frequency range, etc. The vibration absorber tested by vibration table shows that the adjustable frequency range of X, Y, and Z directions are 40, 17, and 50 Hz, respectively. The adaptive frequency adjustment control method designed according to the vibration attenuation characteristics of the piping system can not only ensure the accuracy of frequency adjustment in the entire system operation cycle, but also greatly reduce the need for adaptive frequency adjustment of the vibration absorber due to changes in operating condition time.

(4) According to the typical ship piping system, the adaptive frequency adjustment control system and control method were developed. After single-frequency excitation and experimental verification under actual operating conditions, the vibration absorption effect of the maximum line spectrum of the piping system was 6 dB; the absorption of other line spectra was performed, in which the maximum effect is 9 dB. In the case of reducing the vibration of the piping system, the vibration of the shell can also be effectively reduced, and the control effect of the 250 Hz line spectrum vibration is 15 dB.

Chapter 1
Introduction

Abstract This chapter mainly introduces the necessity of the research and the status of active vibration and noise control, the research status of vibration and noise control of pipeline, as well as the research status of vibration absorption and vibration isolation.

1.1 Background

Sound can travel long distance in water. As a result, attenuating the radiation noise of vessels, i.e., improving their sound stealth, can improve their combat capabilities. At the same time, reducing the vibration of mechanical systems is beneficial to extend the service life span of equipment and offer people aboard a much more comfortable environment. Therefore, researching on vibration and noise control of vessels is of great importance.

The three major noise sources for vessels are mechanical system noise, hydrodynamic noise, and propeller noise. Mechanical system noise includes primary structure-borne sound and secondary structure-borne sound. The former refers to mechanical vibration transmits to shell through internal structure and radiates sound into water. The latter refers to sound, radiates from mechanical chamber to the cabin, and excites the shell vibration then radiates sound into water.

Hydrodynamic noise is sound generated by the pulsation pressure of turbulent boundary layer and other forces acting on the surface of shell. At low speeds, hydrodynamic noise is generally covered by mechanical noise and propeller noise, while becoming the main noise source during mid-high-speed navigation. Propeller noise could be divided into four types: bubble noise, non-uniform flow and uneven flow, and noise caused by the blade, propeller vortex noise as well as propeller turbulence noise caused by unsteady propeller excitation. Among them, vacuole noise is the main noise source. At low speed, propeller noise is not obvious. However, as the speed increases, once the propeller produces cavitation, the noise will suddenly become substantial and become the main noise source of vessels.

F. Wang et al., *Comprehensive Investigation on Active-Passive Hybrid Isolation and Tunable Dynamic Vibration Absorption*, Springer Tracts in Mechanical Engineering, https://doi.org/10.1007/978-981-13-3056-8_1

In summary, mechanical system noise plays a major role when vessels sail at low-mid speed. Controlling and reducing machinery noise are the primary assignment for vessels to be quiet.

As for the mechanical system of vessels, power equipment and piping system are the two main noise sources. Specifically, power equipment can radiate sound through multiple ways as has been aforementioned. Except for radiating noise through outlets directly, piping system can also transmit vibrations to the shell through support elements such as horse feet and pipe clamps to cause shell vibrate.

Traditional passive vibration control methods, such as rubber isolator, steel wire isolator, air spring, constrained damping layer, and floating raft vibration isolation, cannot control low frequency vibration effectively under the condition that volume, weight, and size meet engineering requirements. Active control or semi-active control can achieve targeted control of low frequency vibration since it is based on vibration cancelation or has the ability of structural self-adaptation. It is currently the focus of study of low-frequency line spectrum control.

1.2 Progress of Applied Research on Active Control

Lord Rayleigh wrote in his book, *The Theory of Sound*: Using two electromagnetic synchronization tuning forks to create an interfering sound field, the ear could hear the largest and smallest areas of the volume [1]. This is the first recorded sound field superposition experiment. After that, Coanda [2, 3] and Lueg [4] each independently applied for noise reduction patents based on the principle of sound field interference, in which only Lueg gave a sketch of how to achieve this in the patent application [5]. As a result, in active control domain, Lueg is generally recognized as the first person who reduced noise actively. In addition, Olson conducted a documented first active noise reduction experiment [6, 7] and predicted potential application fields for active noise reduction. However, huge electronic vacuum tubes were not feasible to perform complex signal processing. Therefore, it was not until large-scale integrated circuits became widespread used that active control of vibration and noise began to develop quickly.

According to the control object and control method, active control of vibration and sound could be divided into three aspects: active vibration control (AVC), active noise control (ANC), and active structure acoustic control (ASAC). At present, active control has been widely used in the field of ground transportation, aviation, and navigation. Researches are focused on the study of active control algorithms, novel actuators design and development, and optimal configuration of control systems.

The application of active control in the field of ground transportation mainly focuses on engine vibration and noise control, cabin noise control, seat vibration transmission control, and vehicle suspension vibration transmission control.

Singh et al. [8] analyzed various noise sources of engines. Zhang [9], Olsson [10], Toshio and Itaru [11], Park et al. [12], and Gabbert and Ringwelski [13] studied the issues about engine vibration control.

Stanef et al. [14] and Gulyas et al. [15] separately controlled the cabin noise of mining vehicle and combine harvester. Simulation and experimental results show that significant noise attenuation could be observed at error sensors with adaptive feed-forward control strategy. Geng [16], Guo et al. [17], and Bohn et al. [18] analyzed and studied vehicle vibration caused by ground excitation.

In addition, active control of vibration transmission for car seats is also a research hotspot. Ning et al. [19] designed a novel car seat active suspension system using a low-cost actuator based on a motor and a reducer to control low frequency vibration and large-amplitude vibration from 1 to 4.5 Hz. Results verify the effectiveness of the designed control system. An active seat system proposed by Gan et al. [20] can also greatly reduce the vibration transmitted to the seat and the occupant body under low-frequency periodic excitation. Maciejewski and Krzyżyński [21] studied the simulation of active suspension of vehicle, and experimental results show that the proposed control system design method selected by appropriate controller could convert rigid suspension into soft suspension. In addition, Bianchini [22] introduced an active vibration control system applied to a steering column of a steering wheel aiming to eliminate engine's idle vibration transmitted to the steering wheel.

Active control applications in the aviation field include rotary wing and fixed wing. Researches on the former focus on rotor vibration control and vibration and noise control of cabin. The latter focuses on flutter control and internal vibration and noise control. Miller et al. [23] summarized the current situation of the development of active control systems in rotary-wing machine industry. Zhao and Gu [24], Lu and Gu [25], Sutton et al. [26], Roth et al. [27], and Konstanzer et al. [28] studied rotor vibration control problems of rotary-wing aircraft. In addition, the tonal noise in helicopter cabin has a great negative impact on the psychological and physical health of pilots and passengers. To control this noise, Shenggang et al. [29] developed an active noise control (ANC) system for helicopter cabins. Thomas et al. [30] studied the problem of minimizing vibrational energy of interior acoustic transmission of an aircraft cabin using active control means. Results show that it is difficult to reduce vibrational energy for the frequency of interest significantly. Sun et al. [31], Wang et al. [32], Chen et al. [33], and Zimcik [34] studied the feasibility of using piezoelectric materials to reduce noise in aircraft cabins. Gerner et al. [35] used a step-by-step subtraction method to achieve optimum configuration of actuators and sensors in the active control of noise inside a cabin of military transport aircraft. Experimental and simulation results have good consistency at low frequencies. Griffin et al. [36] described the sensor–actuator architecture used in the definition of an adaptive noise cancellation hardware demonstration using a part of the structure acoustic model of a large aircraft. Karadal et al. [37] and Shevtsov et al. [38] discussed the use of piezoelectric materials to control the tremors of smart fins and rotating propellers, respectively.

The application research of active control in the field of navigation mainly focuses on active control of vibration and noise of the housing, vibration and noise control of diesel-electric system, active control of cabin noise, and active vibration and noise control of shaft. Fischer et al. [39] studied the noise transmission caused by power equipment in ships. Swinbanks and Daley [40], Maillard and Fuller [41], Laplante et al. [42], and Ruzzene and Baz [43] discussed the use of active control to reduce vibration and sound radiation in cylindrical shells. Anand et al. [44] combined a loudspeaker with a hydrophone and mounted them at the bottom of the vessel, as is shown in Fig. 1.1. Then noise radiated from vessel would be reduced. Simulation results show that 20 dB control effect could be obtained in three-dimensional space.

Annaswamy [45] applied active control method to adjust propeller blades as well as related components of underwater crafts to change their basic acoustic characteristics based on bionic concept to reduce acoustic radiation efficiency and improve acoustic stealth performance of underwater vehicles. Pan et al. [46–48], Pan and Hansen [49], and Cao et al. [50] conducted in-depth studies on the vibration and acoustic radiation problems of piezoelectric shells used to control the vibration of cylindrical shells. Caresta and Kessissoglou [51, 52] and Caresta [53] arranged inertial actuators in a circular arrangement, as is shown in Fig. 1.2, to control vibration of submarine shells caused by propeller excitation. Active vibration control and discrete structure acoustic sensing method were researched, respectively. For typical propeller excitation in submarine, simulation results show that only about an actuating force of about 34 N is required to attenuate the sound pressure of submarine shell in concerned frequency range. Annaswamy [45] combined active control and passive control to make underwater vehicle body self-sensing, so that relevant noise characteristics could be adaptively tuned according to environmental and operating conditions. Xu [54] designed a shape memory alloy (SMA) connector to suppress the transmission of vibration waves in a cylindrical shell.

For vibration isolation of ship's power system, it is required that on the one hand, vibration isolation system could support the weight of load without excessive deformation. Specifically, during normal navigation, vibration isolation system should be soft enough to prevent vibration and noise from transmitting through base to shell or

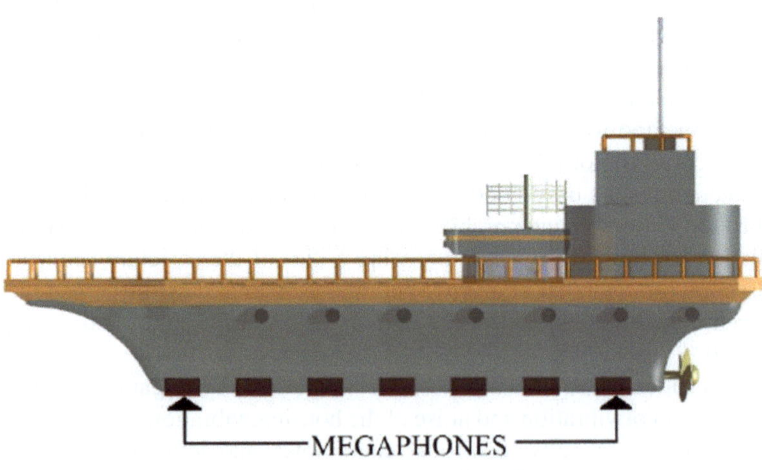

Fig. 1.1 Ship equipped with integrated loudspeakers

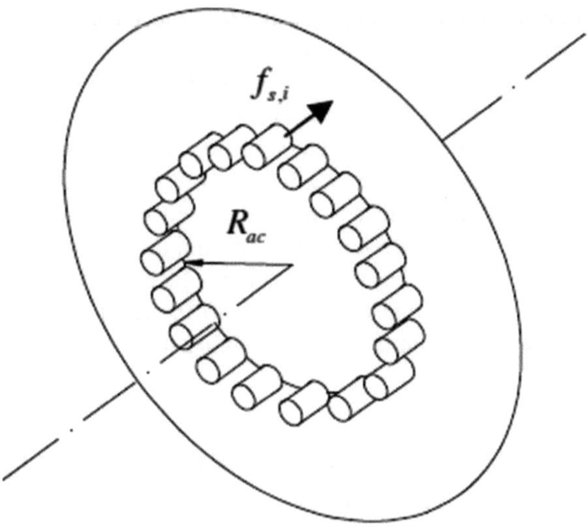

Fig. 1.2 Arrangement of actuators

deck. On the other hand, in severe sea conditions, vibration isolation system should be hard enough that the isolated device could be effectively supported on the housing. In addition, the vibration isolation system can provide enough damping near resonance frequency to reduce resonance peak.

Darsivan and Martono [55] discussed the application of active vibration control in engine vibration and noise control. Zhu et al. [56] developed an active vibration isolation strategy and control system based on adaptive comb filtering for low-frequency vibration characteristics of diesel engines and conducted simulation studies and bench simulations. Olsson [57] studied the problem of active vibration isolation of engine from aspects of engine steady-state and transient internal excitation and object nonlinearity. Johnson and Daley [58, 59] designed an actuator (smart spring) with zero-stiffness characteristic for local displacements of the mounting position below 500 Hz. Experiments show that this kind of actuator has very important application value for reducing vibration transmitted to shell and attenuating the radiation noise. In addition, Daley et al. [60] applied the proposed repetitive control algorithm to active base consisting of "smart spring". The zero-phase shift filter could be used to obtain vibration isolation effect of not less than 50 dB at the target frequency by means of periodic characteristics of the interference signal. Jun et al. [61] studied the feasibility of applying active vibration control technology to the vibration isolation base of Collins-class submarine. Results show that 88, 95, and 97% control effects could be achieved for the engine's rotational frequency, primary and secondary harmonics, respectively. Yang et al. [62, 63], Zhao et al. [64], Niu and Song [65], and Fang and Wang [66] studied the problems of floating raft vibration isolation for naval diesel-electric systems.

In addition, the gearbox is also an important component of the main propulsion system for medium–high-speed diesel engine. With the increase in vibration isolation performance of the diesel engine, the vibration of gearbox is becoming more and more prominent. Zhang et al. [67] and Guan et al. [68] studied the feasibility of controlling gearbox vibration.

Leung [69] wrote in the summary of the feasibility study project of Defense Evaluation and Research Agency (DERA) from 1970 using active control means to reduce ship-borne radiated noise. "The project's research results show that active control measures used to control ship's machinery single-frequency noise is feasible, and the key to success lies in a deep understanding of all possible transmission paths of the noise source. In addition, the purpose of active control is not to replace passive control, but to complement each other. The main obstacle to active control applications is that there are no off the shelf actuators to use. What's more, cost is also an important factor constraining its development."

1.3 Recent Developments of Control Algorithms and Actuators

Control algorithms are common for AVC, ANC, and ASAC. The research includes unconditional and stable feed-forward control. It is necessary to know feedback control of the controlled object model and LMS, RLS, and filter-XLMS algorithms with adaptive adjustment capabilities. Fuller et al. [70], Bies and Hansen [71], Hansen et al. [72], and Elliott [73] have elaborated on these in more detail, and readers can refer to them for a deep understanding.

Current research on control algorithms includes improvements of algorithms and application of algorithms to new control objects.

Phohomsiri et al. [74] and Rohlfing et al. [75] studied the influence of different time delays on feedback control performance in feedback control and the influence of compensation filters on the system performance of feedback control systems. Guo et al. [76] applied differential force feedback control to vibration control of large space structures based on velocity feedback. The simulation results show that the algorithm can provide high damping ratios for the structure, thus suppressing the vibration of structure greatly. Gao and Chen [77] applied nonlinear velocity control with time delay to the active vibration control of bilinear systems. Experimental results show that the feedback gain and delay are two key factors influencing the dynamic characteristics of system and improving the performance of the controller. Martino [78] applied an adaptive filtering algorithm based on fast Fourier transform to control the output force of an electric exciter. Results show that the algorithm can reduce the computational complexity. Perini et al. [79] synthetically used feedback techniques and linear matrix inequality methods to design controllers in discrete state space and applied them to the control of active electromagnetic bearings.

Li [80] used an active linear vibration absorber to apply positive position feedback control to control the large-amplitude vibration of a flexible beam. Results show that positive position feedback strategy could be used to control the large-amplitude vibration of the model. Moreover, if the vibration absorber is properly adjusted with frequency, effective vibration control could be achieved over a wide frequency range. Marx et al. [81] applied the controller based on fast output sampling feedback and periodic output feedback to a piezoelectric structure and obtained an inspiring control effect. Kim et al. [82] used a high-order harmonic LMS algorithm to perform vibration control on a modified 75 cc pump with a swash plate. Both simulations and experiments proved the effectiveness of the control scheme. Mazur and Pawełczyk [83] made full use of the characteristics of rotating devices in a sound insulation cover (the main vibration frequency is fixed) and applied IMC control method to the vibration control of a soundproof cover panel wall. This control scheme belongs to structural acoustic control, for which a very small actuator could reduce substantial radiated noise of device.

Yousefi [84] studied the vibration of flexible structures using piezoelectric sensors and actuators. The purpose was to design a simple and effective active controller for vibration control of flexible structures. Zhang et al. [85] presented a general system control method for the active vibration control of piezoelectric flexible structures, so that the probability information in the parameter uncertainty could be fully studied to ensure the robustness of closed-loop system. The active vibration control system studied by Jovanović [86] includes a strain gauge (sensor), a double-layer PZT piezoelectric actuator, and a composite beam. Research results show that changing the parameters of PID controller can improve the dynamic performance of active control system though it reduces the stability of control system. Meanwhile, the instability of active structure is usually affected by spillover effect.

For ANC, actuators are usually loudspeakers. For AVC, actuators are available in a variety of forms. In early stage, pneumatic, hydraulic, electric, and electromagnetism were all in use. With the rapid development of materials technology, magnetostrictive, ER fluids, magnetorheological fluid, piezoelectric (PZT, PVDF) and MCF and other smart materials are now coming into use. Because of the basic principle of ASAC, acoustic radiation is reduced by controlling the vibration of structure, so that the actuator resembles those in AVC.

Air actuators and hydraulic actuators have unavoidable air leakage and oil leakage drawbacks. What's more, their operating frequency is generally within 10 Hz, which no doubt limits their application area to the control of ultra-low frequency vibration. Electrorheological fluids and magnetorheological fluids are mostly used in semi-active applications for adaptively change the damping of structure. Magnetostrictive actuators have the advantages of large displacement, fast response, high reliability, and low drive voltage. However, a significant hysteresis nonlinearity exists between the applied magnetic field and the output displacement and force of magnetostrictive actuator. Electromagnetic and electric actuators are simple in structure and easy to implement. They have a wide range of applications in engineering practice. Piezoelectric materials have fast response, large output force, and high reliability. Meanwhile, they can sense and actuate at the same time.

Camperi et al. [87] used inertial actuators as is shown in Fig. 1.3 to apply velocity feedback control to reduce the vibration of system under wideband random disturbances. The analysis shows that vibration could be greatly attenuated by increasing the feedback gain to the maximum allowed by system stability. Paulitsch et al. [88]

Fig. 1.3 Inertial actuator

designed a lightweight inertial actuator with an integrated speed sensor for speed feedback control. By arranging the force output and velocity sensor at the same position, theoretically, the unconditional stability of system could be guaranteed. Loussert et al. [89] designed a high-efficiency actuator for active vibration control based on the moving magnet concept. Compared with traditional voice coil type actuator, this new actuator takes advantage in lightweight, larger output force, and required less magnetic field.

Piezoelectric materials are smart materials that have both positive and negative effects and can act as sensors and actuators at the same time. Piezoelectric actuators are widely used for vibration control of structures to improve the performance of controlled systems. Common active control systems based on piezoelectric actuators are composed of piezoelectric actuators, high-voltage power amplifiers, controllers, and corresponding sensors. Monner [90] discussed smart materials that could be used in the field of active vibration noise control, and results show that piezoelectric ceramics should be the primary choice in the field of active vibration and noise control. In the field of early vibration control, research on piezoelectric materials mainly focused on simple structures such as beams, plates, and cylindrical shells. Nelson [91] demonstrated for the first time that piezoelectric shear actuators could be used for active vibration suppression. Sambavekar et al. [92], Kircali et al. [93], Rahman et al. [94], Chhabra et al. [95], Birman [96], Li et al. [97], and Zoric et al. [98] researched on the inclusion of piezoelectric films, in which smart beam modeling and vibration control problems were studied. Yavuz et al. [99] and Berkhoff and Wesslink [100] studied the problem of vibration of plate using piezoelectric plates. Cao et al. [101] used a pair of piezoelectric stack actuators mounted on a housing parallel in the axial direction to drive the actuators of the same phase to control the vibration and acoustic radiation of cylindrical housing with the same magnitude of force. Baillargeon [102] studied active vibration isolation based on PZT reactor actuators. Experimental results show that the "soft base" active vibration isolation system for improving the performance of a passive airbag vibration isolation system could be installed on the STACISTM vibration isolation system developed. The reason is that the magnitude of stiffness of PZT actuator that affects the dynamic characteristics of the control loop and the rigidity of the supporting ground are several orders higher than those of active bladder. Nestorovic et al. [103] studied the use of piezoelectric actuators and sensors to control the vibration of a funnel-shaped structure. Volkan et al. [104] conducted a theoretical analysis of the feasibility of applying active flutter control technology in smart fins. Zhao [105] used two rotary inertial actuators based on piezoelectric materials to control the structural acoustic radiation of rotating equipment. As an additional device, this inertial actuator could be mounted directly on the rotating shaft so that it can intervene as soon as possible into the transmission path of the noise radiation into the noise radiation.

MFC the abbreviation of macro-fiber composite is a novel actuator developed by NASA's Langley Research Center for aerospace applications. It is sandwiched between an adhesive layer and an electrode polyimide film. The rectangular piezoelectric ceramic rods are composed as is shown in Fig. 1.4. Compared with ordinary piezoelectric material has more excellent performance, with a high-cost performance. Sohn et al. [106], Kim et al. [107], and Kumar et al. [108] studied the vibration of shell surfaces using MFC. Williams et al. [109] discussed the production processes and industrial applications of four complex actuators using active fibers: composites, active fiber composites (AFC), MFC, and hollow tube active fiber composites. Leniowska and Mazan [110] studied the use of MFC sensors and actuators to control the vibration of a circular plate. Assuming unknown system parameters, ARC identification method was used to identify the system model through input and output data, and then the resulting linear model was used to use the star. The MFC actuator is used to control the vibration of the circular plate. Both the simulation and experimental results prove the effectiveness of the designed control system.

Brennan and Mcgowan [111] predicted the power demand of piezoelectric actuators used in active vibration control. Calculations and experimental results show that the maximum power required to control the vibration of structure using surface-

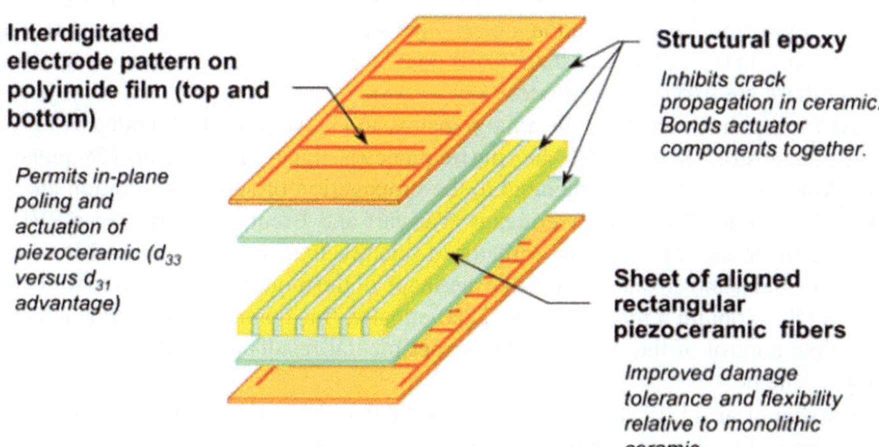

Fig. 1.4 Internal composition of MFC (Smart Materials. Macro Fibre Composites data sheet. Available online: http://www.smart-material.com 03 June, 2008)

contacting piezoelectric actuators is independent of the dynamic characteristics between piezoelectric actuators and accessory structures. For an ideal control system, the required power is a function of the number and type of piezoelectric actuators and the voltage and frequency of the control output signal. Moreover, the required power of the piezoelectric actuator reduces as control efficiency decreases.

1.4 Research Progress of Pipeline Vibration Noise Control

When excited by pulses or mechanical sources, piping system can generate excessive vibrations, which typically include lateral vibration and/or radial vibration of the shell wall. Wachel and Smith [112] discussed several vibration problems in a typical piping system. At low frequencies (wavelengths significantly larger than the diameter of the pipeline), radiation noise is mainly caused by the bending wave of the pipeline. For the right-angle bending in the pipeline, when the plane wave incidents on the elbow, it will produce significant bending wave excitation. Kuhn and Morfey [113] studied transmission losses in long cylindrical steel pipelines. Wachel and Tison [114] discussed the vibration of piping systems containing rotating equipment. Grant [115] studied the critical threshold for fluid instability caused by fluids using the finite element method.

In fluid power systems, the hydraulic pump is one of the causes of pressure pulsation and is also the main cause of hydrodynamic noise generation. This wave, which is generated by the pulsation through pipe wall and fluid, generates a flow-induced vibration and radiates noise out of the piping system. Li and Moore [116] studied noise control of excessive radiation noise during operation of a hydraulic mooring winch system for offshore barges. Pan et al. [117] studied an active valve based on cascading and bypass structures for pulsating pressure control in switched inheritance hydraulic systems (SIHS). Silcox and Elliott [118] used active control to control multi-dimensional random noise in the waveguide. Experiments have shown that a control effect of no less than 20 dB could be achieved for the frequency range where there are two pipeline modes.

As for pipeline vibration, the axial transmission wave (elliptical mode) will produce greater strain in pipe wall when the frequency is higher than the cutoff frequency. This kind of wave is very beneficial to the propagation of acoustic radiation and is difficult to control with passive means. Variyart and Brennan [119] used a PVDF modal sensor and a PZT modal actuator to selectively sense and control this wave. The sinusoidal and cosine shapes (like the shape of the mode shape) are used to shape the PVDF to obtain a modal sensor. Theoretically, this control scheme can achieve complete control of the modality. The experimental results prove the effectiveness of this scheme.

Carsten et al. [120] designed a two-way active dynamic vibration absorber, as is shown in Fig. 1.5, to control the vibration of the pipeline. The advantage of this actuator is that the vibration of the pipeline could be performed over a wide frequency range.

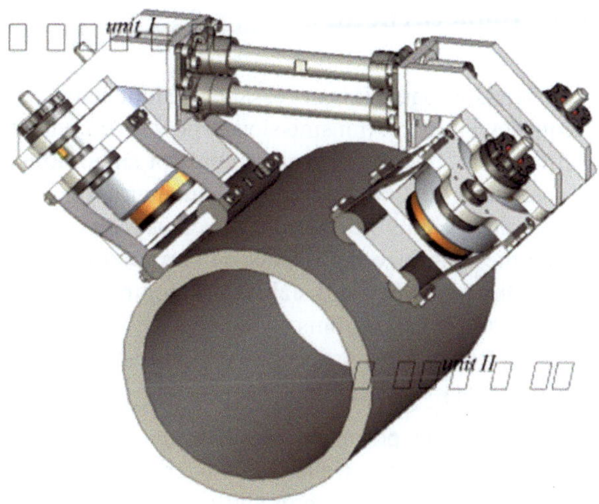

Fig. 1.5 Active dynamic vibration absorber for pipeline

There are four types of transmission waves at the low frequency in the liquid filling pipeline, which are acoustic, longitudinal, torsional, and bending waves, respectively. They weakly coupled to each other, and the fluid and structural responses at low frequencies are dominated by the resonance of each type of wave. Axisymmetric, acoustic, and longitudinal waves dominate the transmission of noise when the frequency is lower than the beam bending mode. Pan [121] studied the piping system excited by a positive displacement pump. Controlled fluid waves are generated using a PZT cylinder actuator embedded in a steel pipe, and longitudinal waves are controlled using three PZT stack actuators arranged along the pipe axis. Earthquake loads in piping systems can cause excessive vibration in the pipeline during an earthquake. Kumar et al. [122] used a semi-active magnetostrictive damper to control the seismic response and used a variety of control strategies to study the damping effect of MR damper. Analytical calculations show that the MR damper was very effective under the optimal parameters. Wang and Sun [123] used an inertial actuator to control the vibration of pipeline. Experiments show that the control effect of not less than 6 dB could be achieved at the shaft frequency and the second harmonic. In addition, at the installation basis for the piping system a control effect of not less than 3 dB could be obtained. Cheer and Daley [124] uses a noninvasive piezo-surrounding piezostack integrated vibration control device to control the vibration and sound radiation of the pipeline. The vibration damping device composed of eight piezoelectric stack actuators uniformly distributed along the radial direction of the pipe. The control strategy is based on an optimal control of time-domain Wiener solution. Kela [125] designed an adaptive Helmholtz resonator and used open-loop and closed-loop control to achieve a control effect of peak-to-peak ripple pressure of not less than 20 dB. Herold [126] discussed the simulation and testing of adaptive Helmholtz resonators. In the frequency range of 100~500 Hz, an average 2.6 dB control effect is achieved, and a 10~18 dB control effect is obtained at the resonance peak.

1.5 Book Structure

As aforementioned, it is necessary to adopt active control means to control the low-frequency vibration transmission of power equipment and piping systems of vessels. Furthermore, due to the requirements of small dimension, light weight, low power consumption and complex environmental applicability, efforts should be made to reduce the weight, the volume as well as the power consumption of control system, meantime minimize the impact on surrounding environment. At the same time, to ensure the reliability and stability of control system, it is necessary to improve the existing control algorithms.

Therefore, for low-frequency vibration transmission control of power equipment, it is necessary to solve:

(1) Improve the power density of actuator and power amplifier, reduce the weight, the dimension as well as the power requirement of actuator and power amplifier;
(2) Study control algorithms and controllers for the characteristics of low-frequency vibration transmission control of power equipment to improve the stability and reliability of control system.

For low-frequency vibration transmission control of piping, it is necessary to solve:

(1) Reduce the size of the overall vibration control structure and reduce the weight and power consumption of the structure;
(2) Study control strategies to improve the control accuracy and reduce the control adjustment time.

Based on the above issues, this book will adopt active–passive hybrid vibration isolation and three-way adjustable frequency vibration absorption technology to control the low-frequency vibration transmission and three-way low frequency vibration of pipelines. The contents are as follows:

The first chapter mainly introduces the necessity of the research and the status of active control, the research status of vibration and noise control of the pipeline, as well as the research status of vibration absorption and vibration isolation.

In the second chapter, based on the Halbach magnetic array, the active and passive hybrid vibration isolation technology for WD618 diesel engine is studied. Firstly, the active and passive hybrid vibration isolator based on Halbach magnetic array is designed. The content includes design requirements, scheme design, and design method.

The third chapter verifies whether the performance index requirements are met and test the resistance and inductance of isolators. Stiffness and static thrust; the vibration isolation effect of the vibration isolator was tested by a single-degree-of-freedom vibration isolation experiment.

The fourth chapter designs and implements the active and passive hybrid vibration isolation control system. The complete active control system includes software and hardware. The software mainly refers to the control strategy and control algorithm implemented in code form. The hardware includes sensors, actuators, power amplifiers, and controllers. This chapter is based on the vibration characteristics and vibration transmission control characteristics of the WD618 marine diesel engine. The hardware design of the active control system includes a controller and a digital power amplifier that satisfies the power requirements of the active and passive hybrid isolator.

The fifth chapter studies the application technology of active and passive hybrid vibration isolation. The active and passive hybrid vibration isolator designed in Chap. 2 and the active control system designed in Chap. 3 are applied to the control of diesel low-frequency vibration transmission. Firstly, the vibration characteristics of diesel engine under typical installation environment and the influence of diesel engine noise on base vibration are analyzed. Based on passive vibration isolation performance test, active control is started to test the effect of active and passive combined control of low-frequency vibration transmission under typical operating conditions.

The sixth chapter studies the three-way adjustable frequency vibration absorption technology of the typical piping system of vessel, proposes the design theory and scheme of the three-way adjustable frequency dynamic vibration absorption, and uses the vibration table to test the vibration performance of the absorber.

The seventh chapter studies and proposes three control methods for different frequency adjustment realization methods. Through debugging comparison, select the optimal control strategy and controller design scheme.

The eighth chapter studies the application of three-way adaptive dynamic vibration absorber in a typical piping system of vessel, including the vibration characteristics analysis of the typical vessel piping system, the vibration absorption effect test under single-frequency excitation, and the vibration absorption effect under the actual working conditions.

References

1. Rayleigh L (1878) The theory of sound, vol. II, chapter XIV, § 282. In: Two sources of like pitch; points of silence; experimental methods, 1st edn. MacMillan & Co, London, pp 104–106; 2nd ed. (1894/96) and Reprints in Dover, New York, pp 116–118
2. Coanda H. Procédé protection contre les bruits. French patent FR 722.274, filed: 21 Oct 1930, patented: 29 Dec 1931
3. Coanda H. Procédé et dispositif de protection contre les bruits. French patent FR 762.121, filed: 31 Dec 1932, patented: 18 Jan 1934
4. Lueg P. Verfahren zur Dämpfung von Schallschwingungen. German patent no. 655 508, filed: 27 Jan 1993, patented: 30 Dec 1937
5. Lueg P. Process of silencing sound oscillations. U.S. patent US 2,043,416, filed: 8 Mar 1934, patented: 9 Jun 1936
6. Olson HF. Electronic sound absorber. U.S. patent US 2,983,790, filed: 30 Apr 1953, patented: 9 May 1961
7. Olson HF (1956) Electronic control of noise, vibration, and reverberation. J Acoust Soc Am 28:966–972
8. Singh A, Bharadwaj S, Narayan S (2016) Analysis of various NHV sources of a combustion engine. Tehnički glasnik 10(1–2):29–37
9. Zhang H (2005) Active dynamic vibration absorber. Harbin Engineering University

10. Olsson C (2002) Active engine vibration isolation using feedback control. Division of Automatic Control, Department of Electrical Engineering, Linköpings universitet, Linköping
11. Toshio Y, Itaru T (2005) Active suspension control of a one-wheel car model using single input rule modules fuzzy reasoning and a disturbance observer. J Zhejiang Univ Sci A 6(4):251–256
12. Park H, Lee BH, Lee CW (2007) Design of an active control engine mount using a direct drive electrodynamic actuator. In: Proceeding of the ASME 2007 international design engineering technical conference & computers and information in engineering conference IDETC/CIE, pp 103–108
13. Gabbert U, Ringwelski S (2014) Active vibration and noise control of a car engine: modeling and experimental validation. In: Mechanics and model-based control of advanced engineering systems. Springer, Vienna, pp 123–135
14. Stanef DA, Hansen CH, Morgans RC (2004) Active control analysis of mining vehicle cabin noise using finite element modelling. J Sound Vib 277(1):277–297
15. Gulyas K, Pinte G, Augusztinovicz F et al (2002) Active noise control in agricultural machines. Proc Int Conf Noise Vib Eng (ISMA) 1:11–22
16. Geng XH (2009) Optimal vibration control for vehicle active suspension system. Ocean University of China
17. Guo NC, Shi WK, Liu WJ et al (2012) Application of dynamic vibration absorber in vibration control of rear axle. J Jilin Univ (Eng Technol Ed) 42(6):1349–1354
18. Bohn C, Cortabarria A, Härtel V et al (2004) Active control of engine-induced vibrations in automotive vehicles using disturbance observer gain scheduling. Control Eng Pract 12(8):1029–1039
19. Ning D, Sun S, Li H et al (2016) Active control of an innovative seat suspension system with acceleration measurement based friction estimation. J Sound Vib 384:28–44
20. Gan Z, Hillis AJ, Darling J (2015) Adaptive control of an active seat for occupant vibration reduction. J Sound Vib 349:39–55
21. Maciejewski I, Krzyżyński T (2011) Control design of semi-active seat suspension systems. J Theor Appl Mech 49(4):1151–1168
22. Bianchini E (2005) Active vibration control of automotive steering wheels. SAE technical paper
23. Miller M, Narkiewicz J, Kania W et al (2006) The application of helicopter rotor blade active control systems for noise and vibration reduction and performance improvement. Pr Inst Lot 1-2(184–185):164–180
24. Zhao CF, Gu ZQ (2009) Time-domain simulation of active control of structural response for helicopter in frequency domain. J Syst Simul 21(20):6347–6351
25. Lu Y, Gu ZQ (2007) Helicopter structure response active control actuator optimization design. J Vib Shock 26(3):23–26
26. Sutton TJ, Elliott SJ, Brennan MJ et al (1997) Active isolation of multiple structural waves on a helicopter gearbox support strut. J Sound Vib 205(1):81–101
27. Roth D, Enenkl B, Dieterich O (2006) Active rotor control by flaps for vibration reduction-full scale demonstrator and first flight test results
28. Konstanzer P, Grunewald M, Janker P et al (2006) Aircraft interior noise reduction through a piezo tunable vibration absorber system. In: Proceedings of the 25th international congress of the aeronautical sciences (ICAS 2006), pp 1–6
29. Shenggang Y, Dakai T, Xiaonei Z et al (2014) Design of active noise control system applied to helicopter cabins. Inter-Noise Noise-Con Congr Conf Proc Inst Noise Control Eng 249(6):1796–1799
30. Thomas DR, Nelson PA, Elliott SJ (1993) Active control of the transmission of sound through a thin cylindrical shell, part I: the minimization of vibrational energy. J Sound Vib 167(1):91–111
31. Sun YF, Chen RW, Xu ZW et al (2003) The active vibration and noise control for fighter cockit model using piezoelectric material. J Astronaut 24(1):43–48
32. Wang HL, Li KX, Chen CL et al (2013) Experimentally investigating active vibration control of panel structure of airplane. Mech Sci Technol Aerosp Eng 32(10):1532–1536

33. Chen RW, Sun YF, Xiong K et al (2003) Structural noise suppression of aircraft cabin using elastic wave control concept. J Nanjing Univ Aeronaut Astronaut 35(5):489–493
34. Zimcik DG (2004) Active control of aircraft cabin noise. National Research Council of Canada Ottawa (Ontario), Institute for Aerospace Research
35. Gerner C, Sachau D, Breitbach H (2004) Active noise control in an aircraft cabin. In: Proceedings of IMAC-XXII, conference on structural dynamics, pp 20040126–20040129
36. Griffin S, Weston A, Anderson J (2013) Adaptive noise cancellation system for low frequency transmission of sound in open fan aircraft. Shock Vib 20(5):989–1000
37. Karadal FM, Nalbantoğlu V, Şahin M et al (2008) Active flutter control of a smart fin. In: 19th international conference on adaptive structures and technologies, Switzerland
38. Shevtsov S, Tsahalis D, Flek M et al (2010) Comparison of active and passive modes of piezoelectric patch actuators for scaled helicopter rotor blade vibration suppression. In: Proceedings of international conference on noise and vibration engineering ISMA 2010, Leuven, Belgium, pp 441–456
39. Fischer R, Boroditsky L, Dempsey R et al (2006) Airborne noise flanking of shipboard vibration isolation systems. Sound Vib 40(12):19–22
40. Swinbanks MA, Daley S (1993) Advanced submarine technology-project M control theory report. Phase 1. GEC-Marconi Research Centre, Chelmsford (United Kingdom)
41. Maillard JP, Fuller CR (1999) Active control of sound radiation from cylinders with piezoelectric actuators and structural acoustic sensing. J Sound Vib 222(3):363–387
42. Laplante W, Chen T, Baz A et al (2002) Active control of vibration and noise radiation from fluid-loaded cylinder using active constrained layer damping. Mod Anal 8(6):877–902
43. Ruzzene M, Baz A (2000) Active/passive control of sound radiation and power flow in fluid-loaded shells. Thin-Walled Struct 38(1):17–42
44. Anand RB, Arun KS, Baskaran K et al (2016) Acoustics reduction in marine vessel using active noise cancellation system. C Int J Eng Adv Res Technol (IJEART) 2(3):22–25
45. Annaswamy AM (2006) Active control of blade tonals in underwater vehicles. Department of Mechanical Engineering, Massachusetts Institute of Technology, Cambridge
46. Pan X, Tso Y, Juniper R (2008) Active control of low-frequency hull-radiated noise. J Sound Vib 313(1):29–45
47. Pan X, Tso Y, Juniper R (2008) Active control of radiated pressure of a submarine hull. J Sound Vib 311(1):224–242
48. Pan X, Tso Y, Juniper R (2005) Active modal control of hull radiated noise. In: Proceedings of acoustics 2005, pp 9–11
49. Pan X, Hansen CH (1997) Active control of vibration transmission in a cylindrical shell. J Sound Vib 203(3):409–434
50. Cao Y, Sun H, An F et al (2012) Active control of low-frequency sound radiation by cylindrical shell with piezoelectric stack force actuators. J Sound Vib 331(11):2471–2484
51. Caresta M, Kessissoglou N (2012) Active control of sound radiated by a submarine hull in axisymmetric vibration using inertial actuators. J Vib Acoust 134(1):011002
52. Caresta M, Kessissoglou NJ (2010) Active suppression of acoustic radiation from a submarine hull using inertial actuators. In: The 20th international congress on acoustics
53. Caresta M (2011) Active control of sound radiated by a submarine in bending vibration. J Sound Vib 330(4):615–624
54. Xu M (2005) Adaptive-passive and active control of vibration and wave propagation in cylindrical shells using smart materials. University of Akron
55. Darsivan FJ, Martono W (2006) Engine mounting characteristic for vibration isolation and active vibration control strategies
56. Zhu MG, Yang TJ, Shuai ZJ et al (2011) Investigation of active vibration isolation based on an adaptive comb-shaped filtered algorithm. J Harbin Eng Univ 32(12):1576–1581
57. Olsson C (2005) Disturbance observer-based automotive engine vibration isolation dealing with non-linear dynamics and transient excitation. Department of Information Technology, Uppsala University

58. Johnson A, Daley S (2011) A smart spring mounting system for marine applications. In: 11th ICSV
59. Daley S, Johnson FA, Pearson JB et al (2004) Active vibration control for marine applications. Control Eng Pract 12(4):465–474
60. Daley S, Hätönen J, Owens DH (2006) Active vibration isolation in a "smart spring" mount using a repetitive control approach. Control Eng Pract 14(9):991–997
61. Li X, Howard CQ, Hansen CH et al (2004) Feasibility of active vibration isolation of diesel engines in Collins class warships. J High Energy Phys
62. Yang T et al (2014) On synchrophasing control of vibration for a floating raft vibration isolation system. In: Inter.Noise 2014
63. Yang TJ, Qi GY, Li WY et al (2006) Study on active control techniques for warship power plant. Ship Sci Technol 28(z2):46–53
64. Zhao YL, He L, Huang YY et al (2005) The computation of shock response of marine floating raft shock-resistant system in the time domain. Noise Vib Control 25(2):14–17
65. Niu JC, Song KJ (2004) Active control strategies of a floating raft isolation system for marine diesel engines. Trans CSICE 22(3):252–256
66. Fang YY, Wang GZ (2006) Design of vibration isolation for ship's auxiliary machinery and analysis of coupling vibration with ship structure. J Jiangsu Univ Sci Technol (Nat Sci Ed) 20(3):16–20
67. Zhang YS, Tong ZP, Zhou Y et al (2013) Research of hard elastic isolation technology of marine gearboxes. Noise Vib Control 3:153–155
68. Guan YH, Shepard WS Jr, Lim TC et al (2004) Experimental analysis of an active vibration control system for gearboxes. Smart Mater Struct 13(5):1230
69. Leung RCN (1997) Active control of machinery noise in a marine environment-lessons learned5. In: Fifth international congress on sound and vibration
70. Fuller CR, Elliott SJ, Nelson PA (1996) Active control of vibration. Elsevier Ltd
71. Bies DA, Hansen CH (2009) Engineering noise control: theory and practice, 4th edn. CRC Press
72. Hansen C et al (2012) Active control of noise and vibration, 2nd edn. CRC Press
73. Elliott SJ (2001) Signal processing for active control. Elsevier Ltd
74. Phohomsiri P et al (2006) Time-delayed positive velocity feedback control design for active control of structures. J Eng Mech 132(6):690–703
75. Rohlfing J et al (2010) Compensation filter for feedback control units with proof-mass electrodynamic actuators. In: Proceedings of ISMA 2010 including USD, pp 425–439
76. Guo T et al (2012) An improved force feedback control algorithm for active tendons. Sensors 12:11360–11371
77. Gao X, Chen Q (2013) Active vibration control for a bilinear system with nonlinear velocity time-delayed feedback. In: Proceedings of the world congress on engineering 2013, vol III
78. Martino OAA (2011) Hybrid time-frequency domain adaptive filtering algorithm for electrodynamic shaker control. J Eng Comput Innov 2(10):191–205
79. Perini EA et al (2009) Active control in rotating machinery using magnetic actuators with linear matrix inequalities (LMI) approach. In: Proceedings of the IMAC-XXVII
80. Jun L (2010) Positive position feedback control for high-amplitude vibration of a flexible beam to a principal resonance excitation. Shock Vib 17(2):187–203
81. Marx LRK et al (2009) Embedded output feedback controllers for piezoelectric actuated structures. World J Mod Simul 5(2):113–119
82. Kim T et al (2016) Active vibration control of axial piston machine using higher harmonic least mean square control of swash plate. In: 10th international fluid power conference
83. Mazur K, Pawełczyk M (2016) Internal model control for a light-weight active noise-reducing casing. Arch Acoust 41(2):315–322
84. Yousefi A (1998) Active vibration control of smart structures using piezoelements. In: CanSmart workshop
85. Zhang K, Scorletti G, Ichchou MN et al (2013) Robust active vibration control of piezoelectric flexible structures using deterministic and probabilistic analysis. J Intell Mater Syst Struct 25(6):665–679

86. Jovanović MM, Simonović AM, Zorić ND et al (2014) Experimental investigation of spillover effect in system of active vibration control. FME Trans 42(4):329–334
87. Camperi S, Ghanchitehrani M, Zilletti M et al (2016) Active vibration control of an inertial actuator subject to broadband excitation 744(1)
88. Paulitsch C et al (2004) Design of a lightweight, electrodynamic, inertial actuator with integrated velocity sensor for active vibration control of a thin lightly-damped panel. In: Proceedings of ISMA 2004
89. Loussert G et al (2016) An efficient and optimal moving magnet actuator for active vibration control. In: 15th international conference on new actuators, Bremen, German
90. Monner HP, Monner HP (2005) Smart materials for active noise and vibration reduction. In: Noise and vibrations—emerging methods
91. Nelson PG (2002) Supporting active electro-pneumatic vibration isolation systems on platforms supported by STACIS TM 'hard-mount' piezoelectric isolation systems
92. Sambavekar RV et al (2015) Active vibration control of a cantilever beam using PZT PATCH (SP-5H). Int J Eng Tech Res (IJETR) 3(5):37–39
93. Kircali OF, Yaman Y, Nalbantoglu V et al (2008) Active vibration control of a smart beam by using a spatial approach. In: New developments in robotics automation and control, pp 1318–1322
94. Rahman, Uralam N, Naushad M (2012) Active vibration control of a piezoelectric beam using PID controller: experimental study. Latin Am J Solids Struct 9(6):657–673
95. Chhabra D, Narwal K, Singh P (2012) Design and analysis of piezoelectric smart beam for active vibration control. Int J Adv Res Technol 1:1–5
96. Birman V (1993) Active control of composite plates using piezoelectric stiffeners. Int J Mech Sci 35(5):387–396
97. Li ZB, Chen H, Zhong YM et al (2010) Experimental research on PPF vibration control of flexible cantilever beam using PZT. J Shenzhen Polytechnic 09(5):1–5
98. Zoric N, Simonovic A, Mitrovic Z et al (2013) Active vibration control of smart composite beams using PSO-optimized self-tuning fuzzy logic controller. J Acoust Soc Am 51(2):275–286
99. Yavuz Y et al (2002) Active vibration control of a smart plate. In: ICAS 2002 congress
100. Berkhoff AP, Wesselink JM (2011) Combined MIMO adaptive and decentralized controllers for broadband active noise and vibration control. Mech Syst Signal Process 25:1702–1714
101. Cao Y et al (2012) Active control of low-frequency sound radiation by cylindrical shell with piezoelectric stack force actuators. J Sound Vib 331:2471–2484
102. Baillargeon BP (2002) Active vibration suppression of smart structures using piezoelectric shear actuators. The University of Maine
103. Nestorovic TT, Köppe H, Gabbert U (2006) Vibration control of a funnel-shaped shell structure with distributed piezoelectric actuators and sensors. Smart Mater Struct 15(4):1119–1132
104. Volkan N, Güçlü S, Ömer FK et al (2008) Active flutter control of a smart fin. In: 19th international conference on adaptive structures and technologies, Ascona, Switzerland
105. Zhao G (2014) Active structural acoustic control of rotating machinery using piezo-based rotating inertial actuators. In: Proceedings of ISMA 2014 including USD 2014
106. Sohn JW et al (2011) Vibration control of smart hull structure with optimally placed piezoelectric composite actuators. Int J Mech Sci 53:647–659
107. Kim HS, Sohn JW, Sohn J, Choi SB (2013) Reduction of the radiating sound of a submerged finite cylindrical shell structure by active vibration control. Sensors 13:2131–2147
108. Kumar GV, Raja S (2012) Sudha V (2012) Finite element analysis and vibration control of a deep composite cylindrical shell using MFC actuators. Smart Mater Res 2090–3561:123–136
109. Williams RB, Park G, Inman DJ et al (2002) An overview of composite actuators with piezoceramic fibers. Proc SPIE Int Soc Opt Eng 4753:421–427
110. Leniowska L, Mazan D (2015) MFC sensors and actuators in active vibration control of the circular plate. Arch Acoust 40(2):257–265
111. Brennan AMC, Mcgowan AMR (1997) Piezoelectric power requirements for active vibration control. Proc SPIE Int Soc Opt Eng 114(9):1542–1570

112. Wachel JC, Smith DR (1991) Vibration troubleshooting of existing piping systems. Engineering Dynamics Incorporated
113. Kuhn GF, Morfey CL (1976) Transmission of low-frequency internal sound through pipe walls. J Sound Vib 47(2):147–161
114. Wachel JC, Tison JD (1987) Vibrations in reciprocating machinery and piping systems. In: Proceeding of the twenty-third turbo machinery symposium
115. Grant I (2006) Flow induced vibrations in pipes, a finite element approach. Nagpur University
116. Li B, Moore S (2014) Noise control for fluid power systems. In: Inter.Noise 2014
117. Pan M, Hillis A, Johnston N (2014) Active control of fluid-bome noise in hydraulic systems using in-series and by-pass structures. In: Ukacc international conference on control, pp 355–360
118. Silcox RJ, Elliott SJ (1990) Active control of multi-dimensional random sound in ducts. NASA
119. Variyart W, Brennan MJ (2004) Active control of the n = 2 axial propagating wave in an infinite in vacuo pipe. Smart Mater Struct 13(1):126–133
120. Carsten B, Jürgen E, Fritz-Otto H (2009) Active control of vibrations in piping systems. In: 20th international conference on structural mechanics in reactor technology
121. Pan X, Forrest JA, Juniper RG (2009) Optimal design of a control actuator for sound attenuation in a piping system excited by a positive displacement pump. In: Proceedings of ACOUSTICS 2009
122. Kumar P, Jangid RS, Reddy GR (2013) Response of piping system with semi-active variable stiffness damper under tri-directional seismic excitation. Int J Struct Eng 258(2):130–143
123. Wang Z, Sun YD (2014) Experimental research on active vibration control of pipe by inertial actuator and adaptive control. J Huaqiao Univ 91(5):725–734
124. Cheer J, Daley S (2015) Broadband active control of noise and vibration in a fluid-filled pipeline using an array of non-intrusive structural actuators. In: Inter-Noise
125. Kela L (2010) Adaptive Helmholtz resonator in a hydraulic system. Int J Mech Aerosp Ind Mechatron Manuf Eng 4(8):684–691
126. Herold S (2012) Noise reduction of a sound field inside a cavity due to an adaptive Helmholtz resonator. In: Proceedings of ISMA 2012-USD 2012, pp 489–504

Chapter 2
Active and Passive Hybrid Vibration Isolation

Abstract In this chapter, based on the Halbach magnetic array, the active and passive hybrid vibration isolation technology for WD618 diesel engine is studied. Firstly, the active and passive hybrid vibration isolators based on Halbach magnetic array is designed. The content includes design requirements, scheme design, and design method; design suitable power amplifier for isolators and verify whether the performance index requirements are met and test the resistance, inductance, stiffness, and static thrust of isolators; the vibration isolation effect of the vibration isolator was tested through a single-degree-of-freedom vibration isolation experiment.

2.1 Preliminaries and Interview

From the principles of passive, semi-active, and active vibration isolation, it could be known that only comprehensive use of passive and active control methods can achieve effective control of vibration transfer in the entire frequency band. From the current realization of active vibration isolation, approaches using airbags or hydraulic are not suitable for applications where the weight or environmental pollution is more demanding due to dimension, weight, post-maintenance, limitations, etc. In addition, due to the low power density characters of electromagnetic actuator, there is also problem that it could hardly meet the actual engineering requirements of dimension and weight.

Halbach array is a new type of magnet structure that combines radial and parallel magnet structure. Provided the end effect is ignored as well as the permeability of the surrounding magnetic material is infinite, a single-side magnetic field will be formed. Compared with traditional permanent magnet motor architecture, Halbach magnets superimpose on each other after the decomposition of parallel and radial magnetic fields. Therefore, the magnetic field strength on the other side is greatly increased, which can effectively reduce the volume of electromagnetic mechanism meanwhile increase the power density of electromagnetic mechanism. What is more, Halbach magnets do not require the armature of magnetic materials to provide access. This not

F. Wang et al., *Comprehensive Investigation on Active-Passive Hybrid Isolation and Tunable Dynamic Vibration Absorption*, Springer Tracts in Mechanical Engineering, https://doi.org/10.1007/978-981-13-3056-8_2

only provides a large selection space for armature materials, but also could greatly reduce the weight of actuator by selecting non-magnetic materials.

Therefore, to increase the output force of actuator meanwhile reduce power consumption, and decrease the dimension and weight, this chapter will design active and passive hybrid isolators based on Halbach magnetic array, including the design method and solution of the isolator, processing and assembling the active vibration isolation unit.

2.2 Vibration Isolation

The vibration power transmitted from machine to base through vibration isolators depends on the dynamic characteristics of base, machine, and vibration isolators. Studying these frequency-dependent characteristics is important for reducing vibration transmission and sound radiation. Transfer impedance method is usually used to describe the dynamic characteristics of isolators; however, this test is only effective in low-frequency regions where isolators exhibit massless spring behavior. The four-pole parametric method could describe the dynamic characteristics of isolators more completely by linking the force and velocity between the input and output. Dickens and Norwood [1] discussed the test procedure of the quadrupole parameter method and the required test equipment in detail and compared it with dynamic stiffness and transfer impedance method. Results show that the measurement of two independent parameters is sufficient to describe the dynamic characteristics of a symmetric isolator. The effectiveness of isolator depends on the four pole parameters of the base and the vibration source and the basic movement characteristics.

It should be noted that to better determine the stiffness of isolator and ensure the success of design, the performance of vibration isolation device needs to be analyzed and evaluated in the vibration isolation design procedure. Then the transmissibility analysis method is generally adopted. However, in this method only the transfer relationship of dynamic forces is considered, while the influence of basic transmission force is not considered, which inevitably causes the deviation between the theoretical prediction and the actual effect to exist. The power flow method studies vibration isolation from the viewpoint of vibrational energy transfer. The power flow contains information on both force and velocity, so that the above problems could be avoided. Douder and White first proposed the concept of [2] power flow. Zhu et al. [3], Wu [4], Xie et al. [5] discussed the use of power flow method theory to evaluate the performance and feasibility of vibration isolation devices in the isolation design of marine engines.

In general, vibration isolation could be divided into passive vibration isolation, semi-active vibration isolation, and active vibration isolation according to vibration isolation methods.

The principle of active and passive hybrid vibration isolation is shown in Fig. 2.1. It consists of three parts: elastic element, damper, and active actuator. Among them, the elastic element is used for carrying the static load of the power equipment on the one hand. On the other hand, it controls the vibration whose frequency is higher

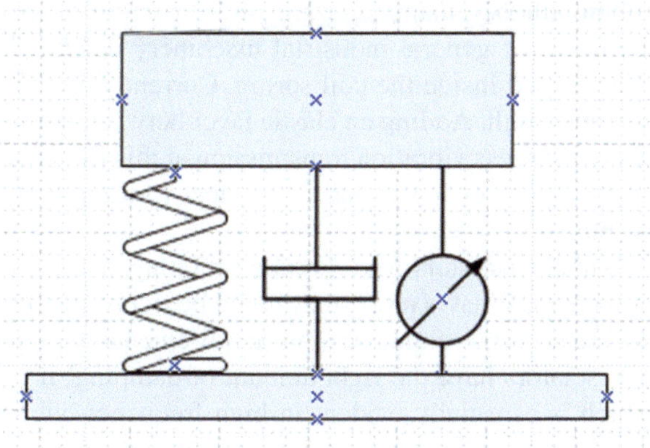

Fig. 2.1 Schematic of active and passive hybrid vibration isolation

than the installation frequency; the damping is used to attenuate the vibration at the resonance frequency. The active element converts electrical energy into mechanical energy. The destructive interference caused by active vibration and disturbing vibration realizes the control of vibration. Control targets are mostly line spectra with prominent amplitudes. As for active control, optional control methods include feed-forward control and feedback control. Feed-forward control is suitable for controlling vibration caused by rotating or reciprocating motion equipment. The control objects include the force transmission and the acceleration of concerned positions.

2.2.1 Passive Vibration Isolation

The commonly used passive isolators are spring isolators, rubber isolators, airbag isolators, and negative-stiffness isolators. Documents [6–10] described the selection and installation methods of vibration isolators in detail. Gu and Zhang [11] briefly explained the installation process and technical requirements of main engine, auxiliary engine, and their base vibration isolators. Song and Li [12] studied the effect of parameter changes in a two-stage vibration isolation system made on the vibration isolation and found that active vibration isolation and passive vibration isolation can use vibration isolation rate to weigh the isolation effect and get the vibration isolation rate, characteristic parameters as well as the relationship between mass ratio, frequency ratio, and relative damping ratio. Sun et al. [13] discussed the setting of unit attributes, meshing, stress concentration in static strength calculation, and modeling methods in the common frame and unit modeling of double-layer vibration isolation systems. Ma and Zhou [14] discussed the technology of float raft isolation applied in the mechanical noise control technology of ships, elaborated on methods, development status, and problems faced by float raft isolation technology.

For spring isolators, the "Spring Design Handbook" systematically written by Jiang [15] introduced concisely the relevant knowledge about materials, design, calculation theory, manufacturing and inspection of springs, which can basically meet the needs of general industrial machinery design. At high frequencies, resonance is generated inside the coil spring. Currently, the vibration transmission efficiency becomes high. Adding an elastic layer between the spring and the support can effectively suppress vibration transmission at this frequency. There is a layer of neoprene on the upper or lower part of product forming coil spring isolator to weaken this problem.

Rubber isolators have good cushioning, vibration isolation, and soundproofing properties. What is more, they could be freely customized for shape and size to meet different stiffness and strength requirements. Compared with spring isolator, rubber isolators have the right amount of damping. It can dissipate vibrational energy, which is especially evident in high-frequency vibrational energy dissipation. The disadvantage of rubber isolators is their performance changes greatly due to temperature changes, and they are also susceptible to corrosion by oil and ozone. Zhao [16], Wang [17], Wu and Dai [18], Shi et al. [19], Zheng [20], and Miao [21] researched on the static stiffness and structural parameters of rubber isolator, as well as the relationship between = material hardness, the development process of the rubber isolator, the aging of the rubber isolator, and the type of rubber isolator was studied independently.

Compared with rubber isolators, metal rubber isolators have good performance in temperature nonlinearity, nonlinear vibration level, and nonlinear inertial overload, especially the creep properties of metal rubber isolator are far less than rubber isolators. Li [22] briefly introduced the characteristics of metallic rubber materials and discussed the nonlinear constitutive relations and research status. Zhao and Liu [23] briefly described the design and calculation methods for steel spring isolators and rubber isolators then introduced several composite isolators based on this.

Richards and Singh [24], Kim et al. [25] and Lin and Lee [26] studied the heuristic constitution equation and finite element equation of the nonlinear characteristics and large static deformation superimposed on small vibration loads of rubber isolators, and the effect that viscoelasticity of rubber makes on isolation performance, respectively. The effects of the formula, the finite element formula, and the viscoelastic properties of the rubber material on the performance of the rubber isolator have been studied and analyzed. The research results of Chavan et al. [27] show that neoprene rubber has better vibration isolation performance than conventional rubber. The static and dynamic characteristics of the 6J X rubber isolators and polyurethane isolators widely used on ships have shown that the polyurethane isolators have higher bearing capacity and wider bearing range than rubber isolators. Vibration reduction effect is good. In addition, studies on the performance of polyurethane isolator with different formulations and hardness show that the overall performance of TDI polyurethane isolator is better.

Compared with the traditional rubber isolator, airbag isolators have advantages of large load (up to 15 t), small structure size, low natural frequency (<5 Hz), no standing wave effect, and high-frequency vibration isolation performance, etc. They could significantly reduce the structural noise of ships, vibration above 10 Hz usually will be isolated by 99%, i.e., the vibration isolation effect is 40 dB. However, because of the internal resonance, the pneumatic system will begin to amplify the vibration from about 1–8 Hz instead of damping; therefore, it also limits the scope of its use to some extent. Xu et al. [28] applied airbag isolator to the vibration isolation of the main engine of the ship. Analysis shows that the airbag isolation system can greatly reduce the transmission of the host excitation force to the hull base. Xiang [29] conducted block-oriented modeling, approximate linearization, and control studies on a nonlinear airbag actuation system. The so-called block-oriented model refers to an input–output model that includes linear dynamic and nonlinear elements connected in series and in parallel. In addition, Chen [30] discussed the feasibility of using the airbag active vibration isolation system to enhance the train ride experience.

There are also some other passive vibration isolation methods, such as wire rope vibration isolator, Zhou and Liu [31] from the perspective of engineering application, the acceleration response characteristics of a steel wire rope vibration isolation system under random load are considered that the system is under certain conditions. It could be described by a linear model. Shu et al. [32] illustrated the nonlinear characteristics of the steel wire rope vibration isolator by means of the vibration isolation design of a large converter unit. Tao [33] tested the static characteristics of the steel wire rope vibration isolator and found that in the y-direction, the bearing and deformation show a soft elasticity relationship. The research of Liu et al. [34] also proved this point.

Mizuno et al. [35] proposed a novel vibration isolation scheme using negative stiffness to achieve infinite stiffness, which uses a negative-stiffness spring in parallel with a common spring and can achieve infinite stiffness using only a simple relative displacement sensor. Greatly attenuate the vibrations transmitted from the ground, as is shown in Fig. 2.2.

Sciulli [36] research shows that the main difference between flexible base rigid equipment (FBRE) and flexible base flexible equipment () systems is the effect of the isolator position on the natural frequency of the system.

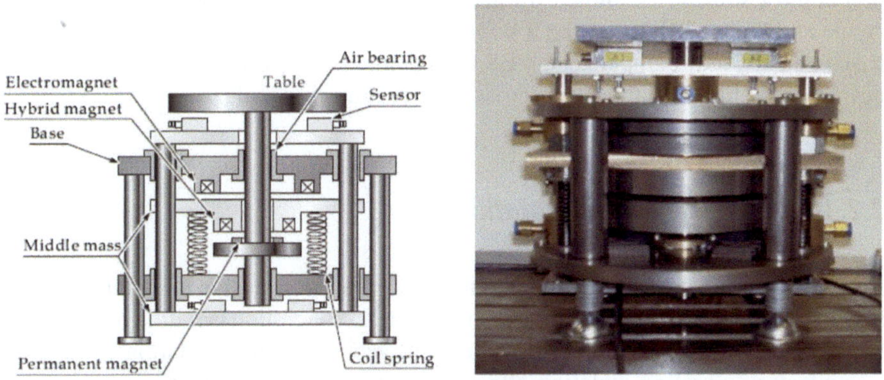

Fig. 2.2 Design draft and prototype of isolator with infinite stiffness

2.2.2 Semi-active Vibration Isolation

Although semi-active vibration isolation is also composed of elastic elements and dampers, unlike passive vibration isolation, it can use controller to adaptively change damping and/or stiffness according to operating conditions. The performance of semi-active control is close to active control, but the cost is lower because only less energy input is needed. However, it should be noted that since its energy requirement is less than that of active control, it cannot achieve the active cancellation of interference, which is also the advantage of active control.

Semi-active vibration isolation two-step design method proposed by Giua et al. [37] is: Firstly, the progressive state estimator is calculated by minimizing the norm of transfer function matrix between the error state estimation and external disturbance and then the target initiative is obtained by solving LQR problem. Finally, the target control law is approximated by controlling the damper coefficient of the semi-active suspension. Shan and He [38] used the principle of controlling the absolute acceleration response by controlling the change of damping ratio in the process of shock response and proposed a two-phase shock isolation semi-active control strategy to achieve no reduction or even increase in impact shock. Under the conditions of isolation efficiency, use the largest possible damping ratio to dissipate the impact energy and reduce the relative displacement amplitude.

Giua et al. [37] proposed a design method for semi-active suspension system. Firstly, consider a target active control law in the form of feedback control law. Secondly, the target control law is approximated by controlling the damping coefficient of the semi-active suspension system. Two different types of shock absorbers have been studied in particular: The first is the use of magnetorheological fluids instead of oil, and the second is a solenoid valve shock absorber. Maciejewski and Krzyżyński [39] discussed the controller design for semi-active seat suspension. Based on the inverse dynamic characteristics of spring and damper elements, a semi-active vibration control strategy is studied. In addition, to reduce the radiated noise of ships, Ahuja and Gupta [40] proposed a semi-active control scheme for variable damping based on buoyancy dampers using ER fluid dampers. The fuzzy logic controller was designed by analyzing the characteristics of excitation signals.

2.2.3 Active Vibration Isolation

Bing et al. [41] analyzed common vibration sources and vibration isolation measures in aviation and navigation. In terms of theory, passive vibration isolation measures for low frequency vibration are not suitable while active vibration isolation can achieve a satisfactory vibration isolation effect at low frequencies. Furthermore, the active damping system has no resonance and no vibration is amplified at any frequency.

The active vibration isolation system basically consists of a passive vibration isolation element, an actuator, and a control system. According to the combination form of actuator and passive isolation element, the structure of active isolation could be divided into parallel and series, as are shown in Figs. 2.3 and 2.4, respectively. In parallel vibration isolation, the actuator directly acts on the controlled object. It is used to control the vibration that passive vibration isolation cannot isolate and is suitable for controlling the vibration of the load. Series vibration isolation cannot directly act on the load, and thus it does not sense the resonance of the load and is suitable for controlling the vibration of ground. Especially when the piezoelectric actuator is used as a main action moving element, it is more suitable for controlling the low-frequency ground vibration.

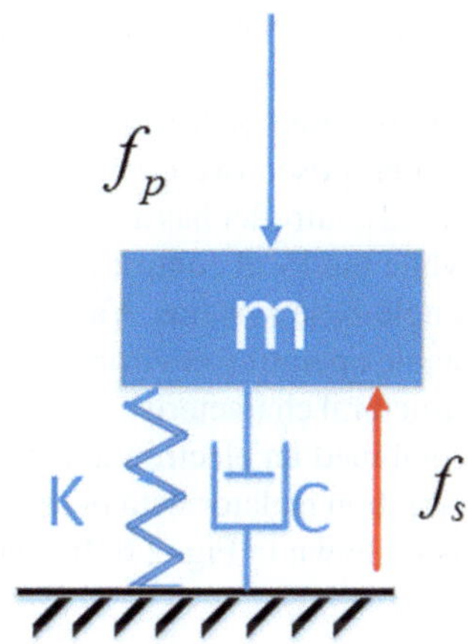

Fig. 2.3 Vibration isolation system with actuator and passive vibration isolation elements arranged in parallel

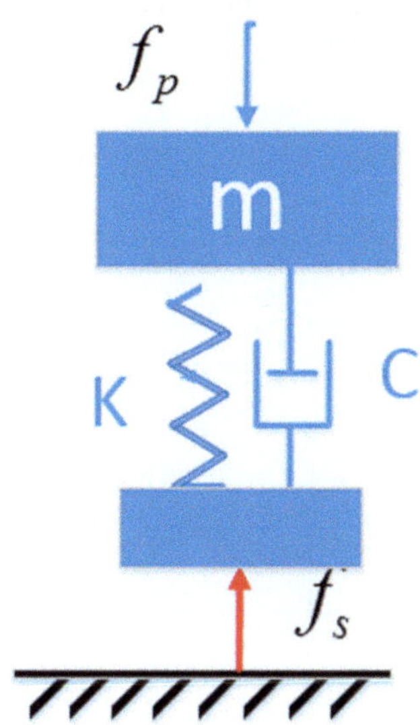

Fig. 2.4 Vibration isolation system with actuator and passive vibration isolation elements arranged in series

Researches on active vibration isolation could be divided into new vibration isolator design, vibration isolation strategies study, and attempt in applying active vibration isolation methods into new fields.

Jian et al. [42] briefly introduced the application of active and passive hybrid vibration isolation devices in the field of vibration control of ships and equipment. Huang [43] illustrated the asymptotic stability of the vibration of the control system and the damping mechanism of the control system based on the relationship between the energy distribution change and the state transition motion of the AVS structure control system at the instant of the variable structure intuitively. Long et al. [44] analyzed the mechanism and characteristics of magnetic levitation vibration isolation and showed that the idea of magnetic levitation vibration isolation is feasible. The performance of vibration isolation depends on many factors such as the structure of the electromagnet, power consumption, and suspension clearance. Hoque et al. [13] developed a three-degree-of-freedom active vibration isolation system with zero-power controller and proposed two control strategies. The experimental results show that the model-based controller has a good effect on the control of multi-degree-of-freedom systems, while the local control is more suitable for controlling single-degree-of-freedom or single-base systems. Yang et al. [45] designed an oil damping isolator to provide a stable operating environment for precision equipment and studied and optimized the structural characteristics from both theoretical and experimental points. He et al. [46] combined an electromagnetic actuator (Fig. 2.5) with an airbag isolator designed a vibration isolator with both active and passive hybrid vibration isolation capabilities, as is shown in Fig. 2.6. It is applied to low-frequency vibration transmission control

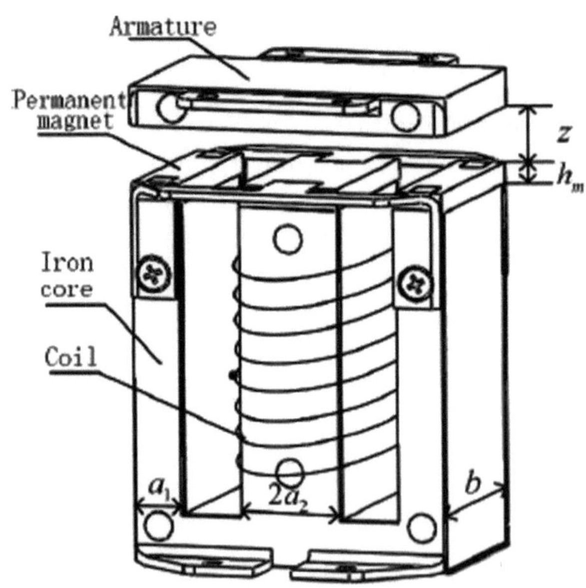

Fig. 2.5 Electromagnetic actuator

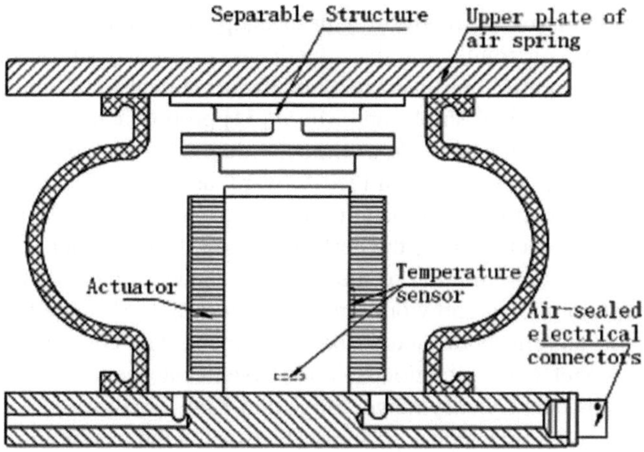

Fig. 2.6 Active and passive hybrid vibration isolator integrated with electromagnetic actuator and airbag

of a diesel engine, and a significant vibration reduction effect is obtained for line spectrum below 200 Hz.

Hoque et al. [47] designed a magnetically suspended active vibration isolator based on zero-power demand magnetic suspension vibration isolation technology and applied it to microvibration isolation control. The experimental results showed that both static and dynamic responses to direct interference and the isolation effect to ground vibration are ideal.

Researches on active vibration isolation strategy are: Lin and McInroy [48] combined adaptive sine-wave interference cancellation control method and Stewart platform fault-tolerant pointing algorithm to obtain a fault-tolerant pointing control strategy capable of low-frequency tracking, realizing large single active interference suppression in low frequency band and passive vibration isolation in high frequency range. El-Sinawi [49] applied active vibration isolation to control the vibration of the flexible beam installed on the basis of vibration and elasticity and adopted an active feed-forward and feedback control method based on Kalman estimator to reduce the force transmitted from the foundation to the structure. Baig and Pugazhenthi [50] used genetic algorithm (GA) training neural network to optimize the design parameters of active vibration Stewart platform (SP) and studied the influence of design parameters on the optimization. Liu et al. [51] proposed a new active vibration isolation control method based on adaptive notch filter that can effectively isolate periodic vibrations. Chen [52] established a theoretical model for a float raft vibration isolation system with an active dynamic absorber by the subsystem admittance synthesis method.

In the new field of application of active vibration isolation methods are: Li [53] using Newton-Euler method to establish the dynamic model of a 6RSS parallel mechanism of the main body of the six-dimensional damping platform, from the perspective of synchronization and optimization of control and institutions. The parameters of the vibration-reducing platform with 6RSS parallel robot as the main body are optimized. Through simulation, the effectiveness of the synchronization and optimization of the control and main body of the six-dimensional vibration-reduction platform is illustrated. Yoshioka and Murai [54] proposed an active control system that considers the bending mode of the vibration isolation table and uses absolute velocity feedback, ground motion feed-forward, and virtual TMD methods to control the rigid body motion and bending modes of the vibration isolation table. The genetic algorithm can improve the performance of the controller. The experimental results show that the vibration of the vibration isolation table could be attenuated to one percent of the ground vibration. Thorsten et al. [55] studied the active vibration isolation table in both theory and experiment. The control results show that the multi-channel active vibration isolation system can suppress the vibration of the structure in a wide frequency range. Singh and Kim [56] discuss vibration isolation measures for multidimensional systems. The isolator and receiver are modeled using continuous system theory. Aso [57] proposed a so-called suspension point interferometer (SPI) active vibration isolation scheme to improve the sensitivity of detecting the weak influence of gravitational waves. The test results show that the MIF could be implemented within the frequency range of 10 Hz and below. Noise is reduced by 40 dB. Arias-Montiel et al. [58] studied the modeling, analysis, and unbalanced response control of systems with dual disks. The experimental results show that the unbalanced response of the first disk could be reduced by 66% and the second by 44%.

For noise, vibration, and harshness (NVH) in the field of passenger tools, Ahn [59] designed a vibration isolation system using negative-stiffness vibration isolation principle and applied it to a vibration isolator of a seat in combination with an airbag isolator. Fredrik and Oskar [60] proposed a parametric model that is valid in the frequency range below 300 Hz. The calculated results are in good agreement with the experimental results. Hassan [61] discussed the damping performance of passive, semi-active, and active suspension systems under realistic road excitation based on a simple vehicle model. Balossini et al. [62] considered using a hydraulic actuator to generate compliance with appropriate control strategies so that it could be used for active lateral support of high-speed trains.

2.3 Control Plant and Vibration Characteristics Analysis

2.3.1 Control Plant

The WD618 marine diesel engine (as is shown in Fig. 2.7) was selected as the control object of vibration isolation. The parameters of the diesel engine are: rated speed 1500 r/min, rated power 220 kW, idle speed 650 r/min. The air intake method is pressurization and cooling, and it is a water-cooled, in-line, four-stroke type. The number of cylinders is six, the cylinder diameter is 126 mm, the piston stroke is

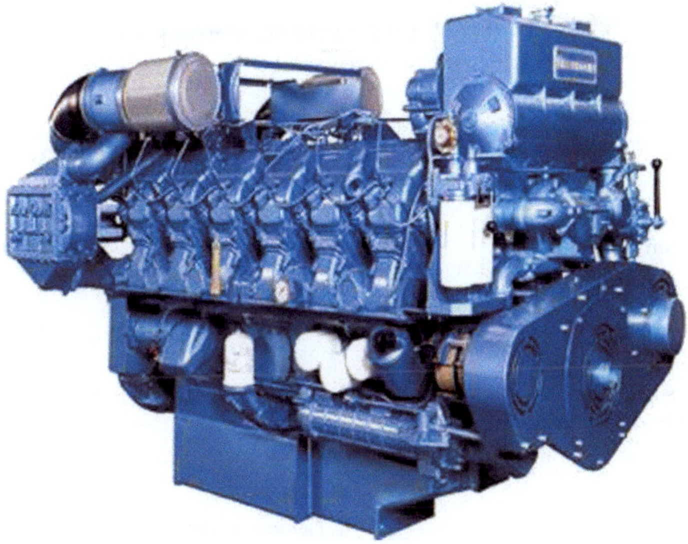

Fig. 2.7 WD618 marine diesel engine

155 mm, and water-cooled forced circulation is adopted. Displacement 11.596L, net weight 1100 kg, dimension 1488 mm × 872 mm × 1258 mm.

According to the weight of diesel engine and vibration acceleration and displacement under rated conditions, the design index of the active and passive hybrid vibration isolators could be roughly quantified as:

(1) The static load that a single vibration isolator can withstand should be not less than 275 kg.
(2) The main action power is not less than 1500 N ($1100\,\text{kg} * 1.2\,\text{m/s}^2$).
(3) The activation stroke is not less than 1 mm ($1.2/200^2$, the maximum frequency of active control is 200 Hz).

The design procedure is: through testing and analyzing the vibration characteristics under typical working conditions of the diesel engine, quantify the required power and stroke of the actuator, determine the installation frequency of active and passive hybrid vibration isolators, and calculate the required spring stiffness, the amount of magnet, number of coil turns, and appropriate operating current required through calculating the power and actuating stroke; and considering the requirements of heat dissipation and installation methods, the possible structures of active and passive hybrid vibration isolators are designed; after assembly, test the performance parameters and further test the vibration isolation performance, including single-frequency vibration isolation performance test and vibration isolation performance test under practical application environment. The specific flow chart is shown in Fig. 2.8.

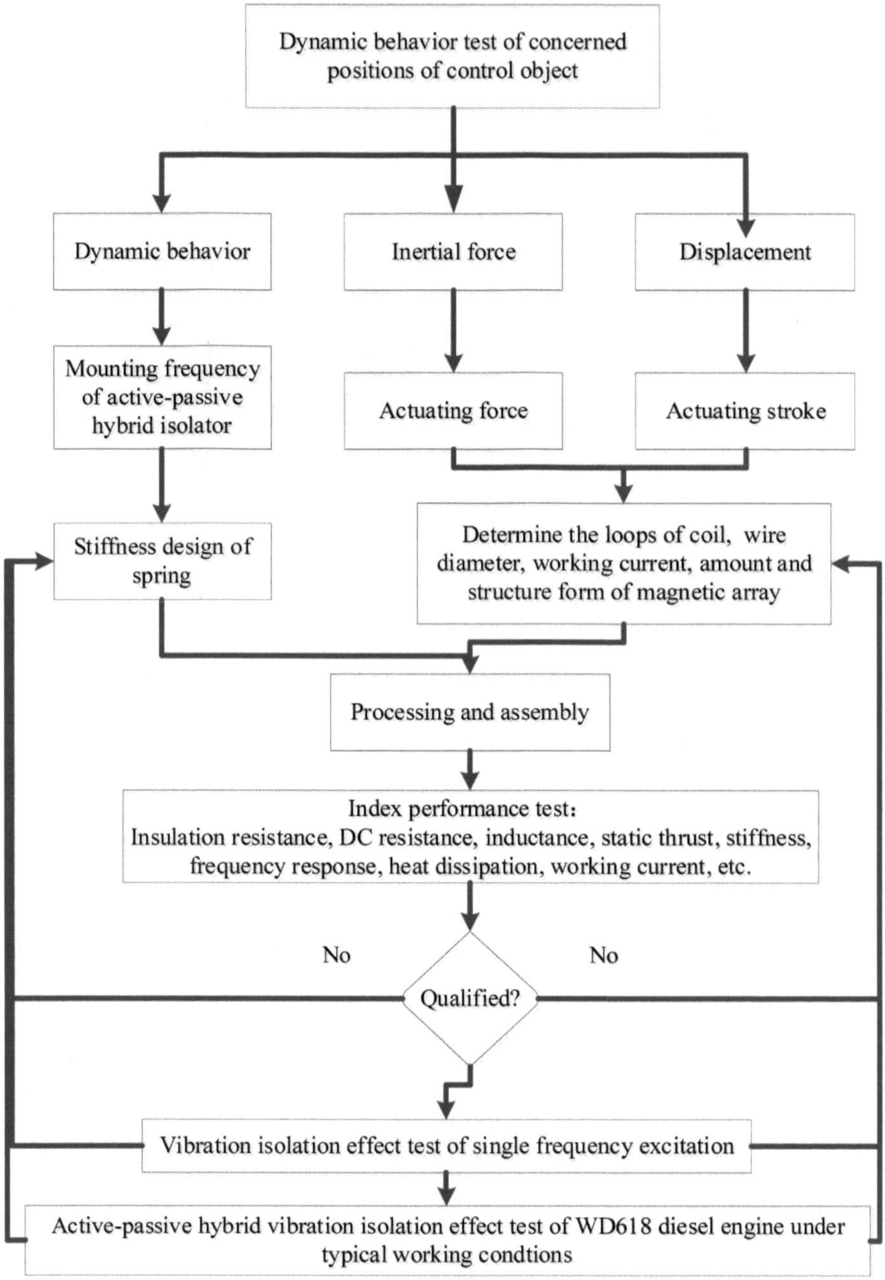

Fig. 2.8 Design framework for active and passive hybrid vibration isolators

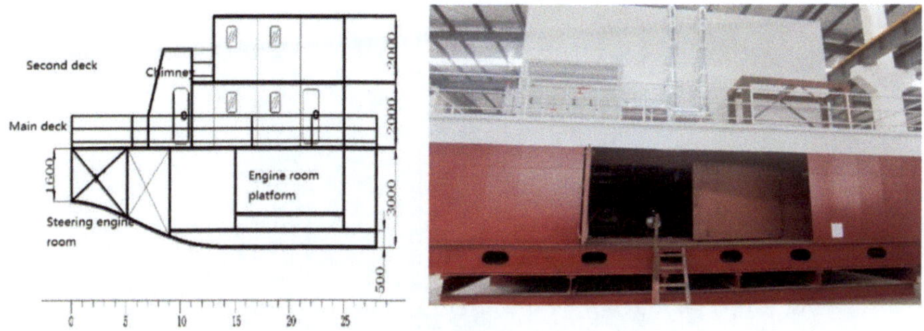

Fig. 2.9 Schematic of large-scale model (left) and spot scene (right)

2.3.2 *Test Environment*

The control object WD618 marine diesel engine is installed in a large-scale model, the overall model is placed flat on the ground, and the model simulates the actual civil ship part line type, including the engine room area (including diesel engine, centralized control room, and workroom), main deck (including four cabins and air-conditioning), dimethyl board (including four cabins) equipped with cabin fan vents. Ventilation piping system is arranged in the engine room area, and the air-conditioning and ventilation system are arranged on the main deck and the deck board. The models are all steel structures, and no sound-absorbing and sound-absorbing materials are used on bulkheads, decks, and cabins, as is shown in Fig. 2.9.

To compare the vibration isolation effects of the active and passive hybrid vibration isolation, the vibration characteristics of the diesel engine, the foot, the base, and the deck under a typical installation environment were tested and analyzed.

The WH400 passive isolator of Wuxi Shenjie Shock Absorber Factory was selected. The diesel engine was connected to the isolator via a 15 mm transition plate. The plane connecting the isolator to the base was about 45° from the ground. The thickness of the base plate was 25 mm. The two vibration isolators are mounted on the stern of the diesel engine and are mounted on the same base. The stern base and the stern are the same. The overall installation is shown in Fig. 2.10.

2.3.3 *Vibration Characteristics Analysis*

Through the test could be found [63]:

(1) The 1/3 octave spectrum of the longitudinal vibration of the diesel engine's feet has a peak at 125 and 1000 Hz. The rated speed is 1500 r/min and 125 Hz reaches 130 dB, 1000 Hz reaches 140 dB, and the total vibration level from 10 to 6400 Hz is 146.4 dB. The vertical vibration of 125 Hz on the base panel

Fig. 2.10 Arrangement of WD618 diesel engine

below the isolator has the largest vibration level. The 10–125 Hz rises. The 200 Hz high-frequency wave fluctuates around 90–100 dB. There is a small peak at 1000 Hz. The 125 Hz peak at the rated speed of 1500 r/min 120 dB is reached. From 700 to 1100 r/min, the longitudinal vibration isolation effect of the isolator is about 20 dB, 1300–1500 r/min, and the vertical vibration isolation effect of the isolator is about 16 dB, as is shown in Figs. 2.11 and 2.12.

(2) In addition to the idle speed of the diesel engine's feet, the vibration does not increase significantly with the change of the rotational speed. The 10–1000 Hz is in an upward trend. There is no obvious peak frequency in this frequency band. The total lateral vibration level of the machine feet is 148.7 dB at a rated speed of 1500 r/min. The lateral vibration of the panel below the isolator exhibits double-peak characteristics, with large peaks at 125 and 1000 Hz, and a large vibration level at 1000 Hz centered frequency bands (500–1250 Hz). The rated magnitude at the two peak frequencies is approximately 110 dB, and the total lateral vibration level at the rated rotational speed of 1500 r/min is 116.7 dB. The lateral vibration isolation effect of the isolators exceeds 30 dB at different rotational speeds, as is shown in Figs. 2.13 and 2.14.

(3) The vertical vibration of the diesel engine's feet is the same as the lateral vibration of the feet. It reaches a maximum peak of more than 140 dB at 1000 Hz, and the total vertical vibration level of the foot is 150 dB at a rated speed of 1500 r/min. The vertical vibration of the base plate under the isolator is the same as the lateral vibration of the base. It also exhibits a double-peak characteristic. The main vibration levels are mainly contributed by the frequency band

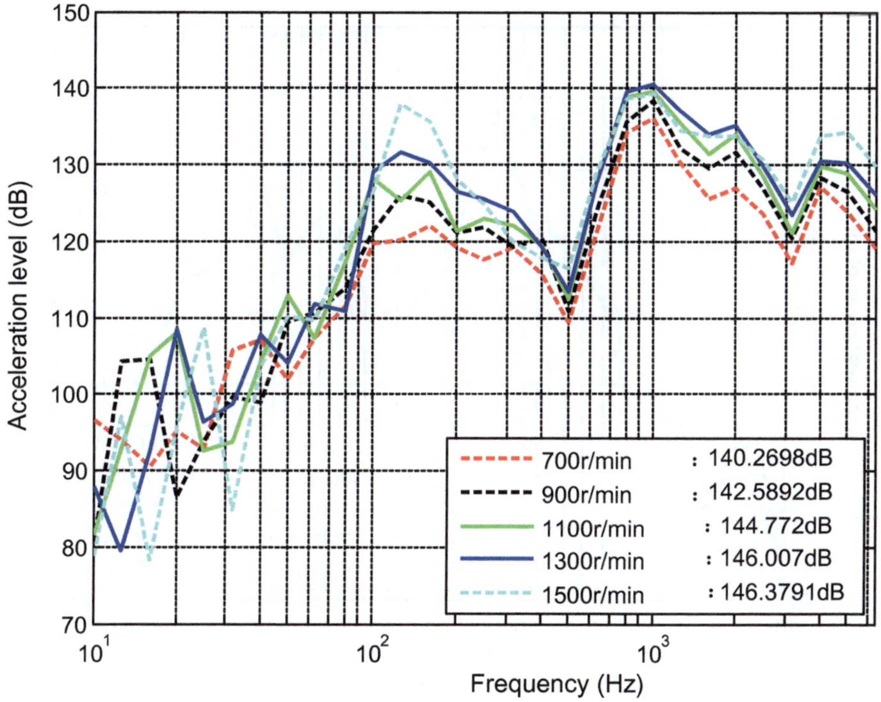

Fig. 2.11 Longitudinal vibration levels of feet for different rotating speeds

range from 125 to 1000 Hz, and the magnitudes of the two peak frequencies at the rated rotation speed exceeded 110 dB. The total vertical vibration level of base at the rated speed of 1500 r/min is 123.4 dB. The vertical vibration of deck below the susceptor faceplate is almost the same as that of the susceptor faceplate. The vertical isolation effect of isolators at different rotational speeds is approximately 27–30 dB, as is shown in Figs. 2.15, 2.16, and 2.17.

It could be observed that the base and deck vibrations are prominent before 200 Hz, and the single-line spectrum peaks between 80 and 150 Hz are obvious, such as 88, 100, 125, 138, 150 Hz, which are the frequencies that should be actively focused on.

2.4 Design Scheme

In 1979, when the American scholar Klaus Halbach did electron acceleration experiments, he discovered the special permanent magnet structure of Halbach array and gradually perfected this structure, eventually forming the so-called Halbach magnetic array.

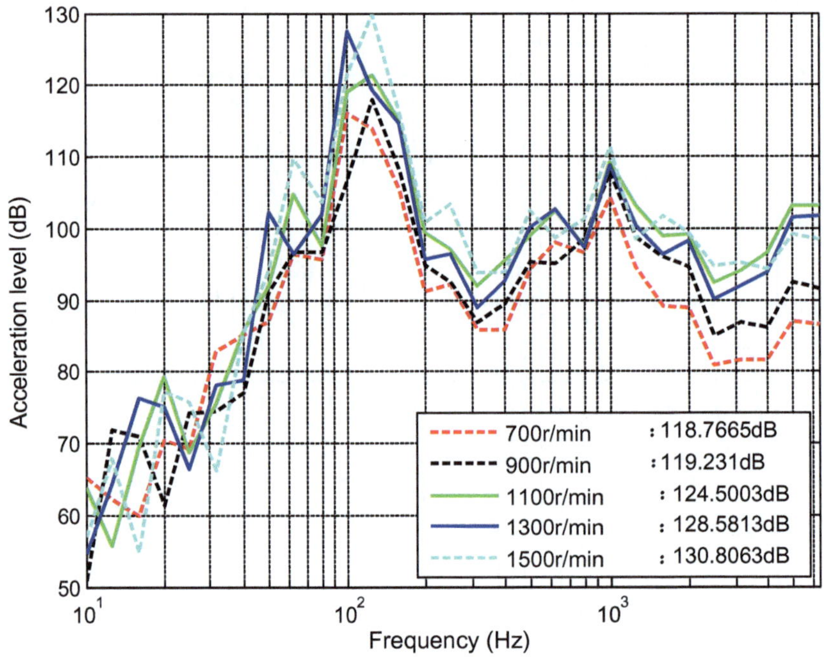

Fig. 2.12 Longitudinal vibration levels of bases for different rotating speeds

The Halbach magnetic array combines the radial and parallel arrangement of the magnets. If the end effect is ignored and the permeability of the surrounding magnetic material is regarded as infinity, the permanent magnet structure eventually forms a one-sided magnetic field. Therefore, compared with the traditional permanent magnets, the parallel magnetic field and radial magnetic field after the decomposition of the Halbach magnetic array are superimposed on one another, so that the magnetic field intensity on the other side is greatly increased, so the power density of the magnet could be significantly increased, and then the number of magnets could be reduced. In addition, since the unilateral magnetic field distribution generated by the self shielding effect of the Halbach magnetic array no longer requires the armature to provide a path for the magnetic material; therefore, applying this magnetic array not only provides a large amount choices for the selection of the armature material; but also, based on the choice of non-magnetic material could reduces the weight of actuator. Material also reduces the weight of the structure.

Therefore, designing an electromagnetic actuator based on this magnet structure can greatly reduce the volume and weight of the actuator while ensuring effective actuating force, and the high-power density characteristic of the Halbach magnetic array can substantially reduce the power consumption that actuator demanded.

Refer to the active suspension device designed by Sande [64]. The design schematic of the active and passive hybrid vibration isolators is shown in Fig. 2.18.

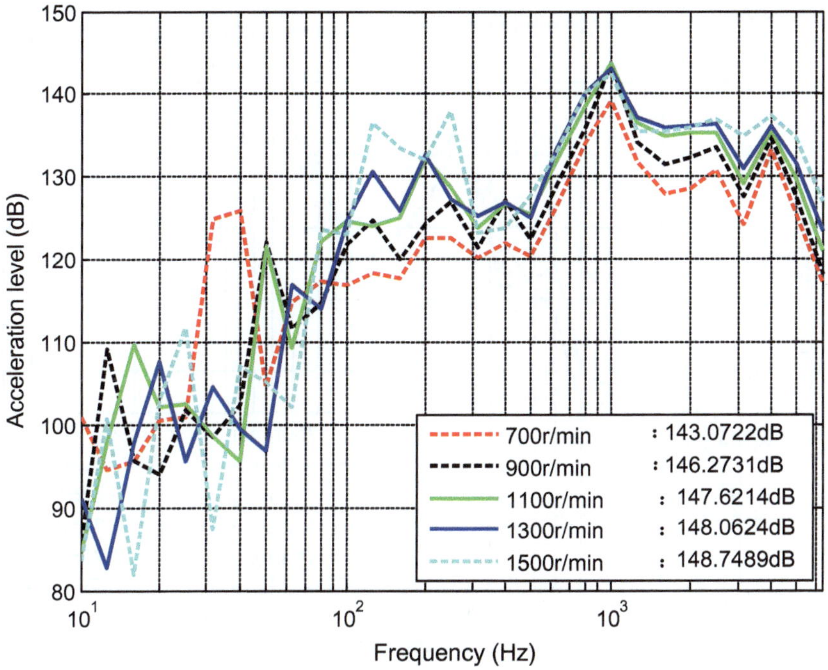

Fig. 2.13 Lateral vibration levels of feet for different rotating speeds

The main components are: 1 is a connecting plate, 2 and 4 are displacement sensors, and 3 is a passive damping element. 5 is a cylinder block, 6 is a temperature sensor, 7 is a Halbach magnetic array, 8 is a slide bearing, 9 is a hollow cylinder, 10 is an excitation coil, 11 is a hydraulic oil that provides damping and heat dissipation, and 12 is a buffer. Based on the working principle of active and passive hybrid vibration isolators of the Halbach magnetic array, the coil spring acts as an elastic element and acts as an ordinary vibration isolator. When the load changes, the controller processes the vibration signal collected by acceleration sensor, and after processing and calculation, the input current of the armature is controlled to achieve adaptive control of the vibration. In addition, the magnetic field generated by cutting the excitation coil/permanent magnet is moved up and down by the armature, and the generated electromagnetic force is opposite to the direction of vibration and acts like an eddy current damper. Hydraulic oil, on the one hand, can dissipate heat to ensure that the actuator will not overheat under high frequency response; on the other hand, it acts as a damping material to relieve severe vibration and shock.

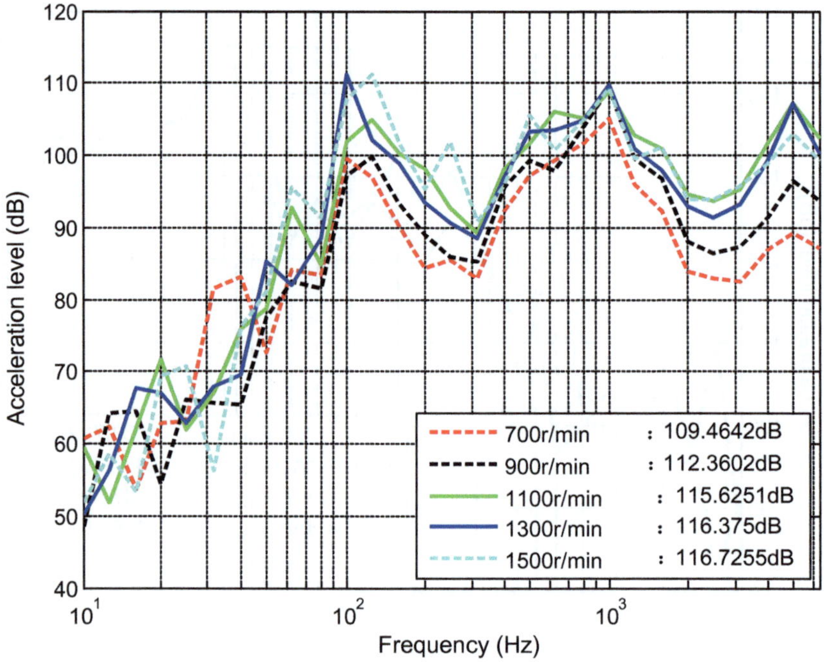

Fig. 2.14 Lateral vibration levels of bases for different rotating speeds

2.4.1 Circuit Model

The model of the motor part of the isolator is composed of voltage source V_i, resistance R_i, inductance L_i, and back electromotive force E_i, as are shown in Fig. 2.19, thus characterizing the differential equation of this circuit as:

$$V_i = E_i + R_i I_i + L_i \frac{\mathrm{d}I_i}{\mathrm{d}t} \tag{2.1}$$

where subscript i values a, b, and c represent three phases, R_i and L_i respectively represent the resistance and inductance of each phase. The current in each phase are:

$$I_a = \hat{i}\sin\left(\frac{\pi z}{\tau_p} + \phi\right) \tag{2.2}$$

$$I_b = \hat{i}\sin\left(\frac{\pi z}{\tau_p} - \frac{2\pi}{3} + \phi\right) \tag{2.3}$$

$$I_c = \hat{i}\sin\left(\frac{\pi z}{\tau_p} - \frac{4\pi}{3} + \phi\right) \tag{2.4}$$

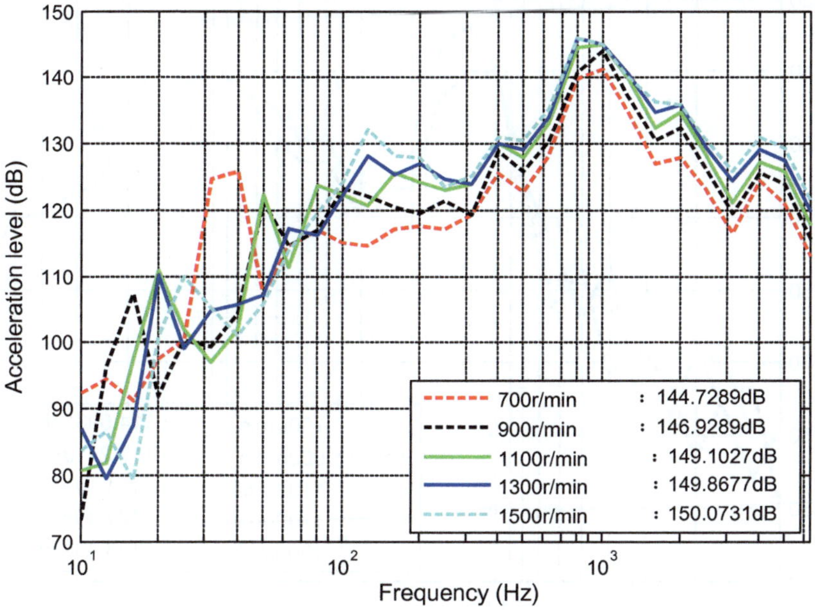

Fig. 2.15 Vertical vibration levels of feet for different rotating speeds

where $\hat{i}$ is the magnitude of the current, z is the displacement of the actuator, τ_p is the pole pitch, and ϕ is the initial phase.

The definition of back electromotive force is

$$E_i = k_{Ei} v \tag{2.5}$$

where k_{Ei} is the gain of back electromotive force and v represents the actuator's operating speed. Assuming $\hat{i}$ and v are independent, the driving force of the actuator is:

$$F_{act} = F_{current} + F_{damp} = k_I \hat{i} + dv \tag{2.6}$$

That is, the total working force is the resultant force of the electromagnetic force and the damping force, where k_I is the force amplification gain, d is the damping coefficient and could be determined by testing.

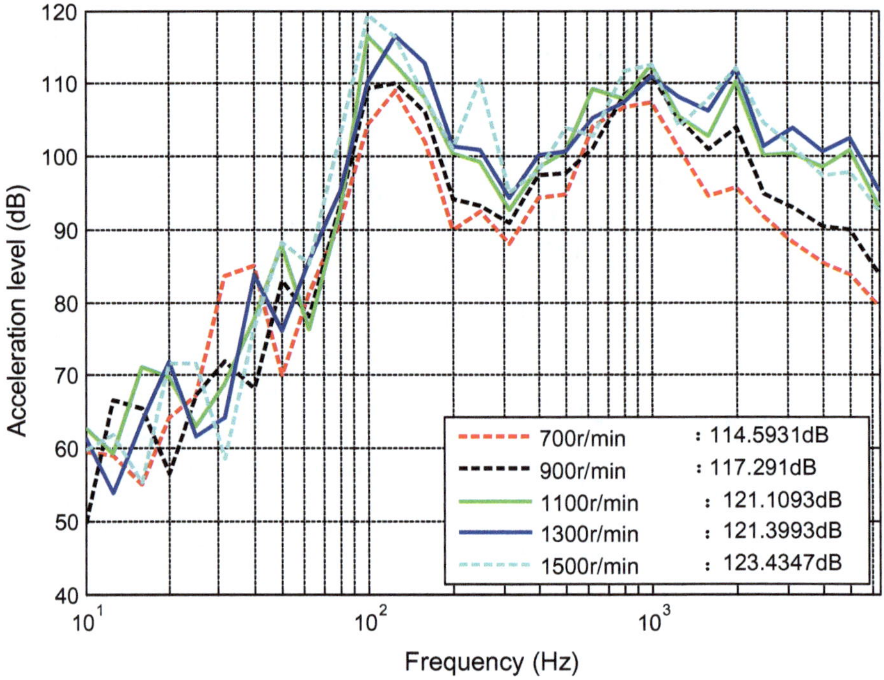

Fig. 2.16 Vertical vibration levels of bases for different rotating speeds

2.4.2 *Spring Stiffness*

Considering the natural frequency of the system after diesel engine isolators are installed:

$$f = \frac{1}{2\pi}\sqrt{\frac{K}{M}} \tag{2.7}$$

Thus, the spring rate is $K = (f \times 2\pi)^2 \times M$.

It is required that this frequency should be as low as possible based on the frequency of 25 Hz (1500 RPM) corresponding to the normal operation of diesel engine. Substituting the mass of the diesel engine $M = 1100\,\text{kg}$ into Eq. (2.7), the upper limit of the stiffness of a single vibration isolator could be obtained, i.e.,

$$k \leq 4.3382 \times 10^4 \times f \tag{2.8}$$

In addition, consider the maximum allowable spring deformation of structure, we can get the lower limit of the stiffness, if the allowable deformation is Δl, then

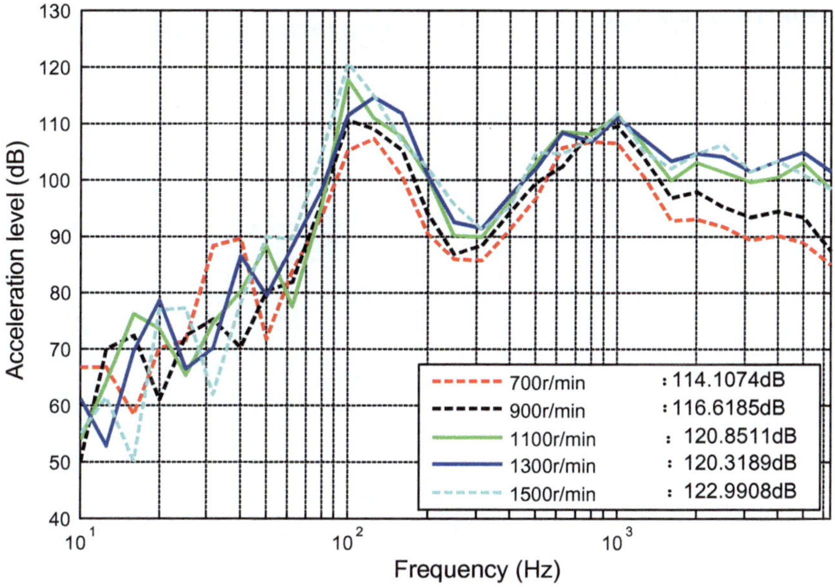

Fig. 2.17 Vertical vibration levels of diesel engine for different rotating speeds

$$k = \frac{F}{\Delta l} \geq \frac{M \times g}{4 \times \Delta l} = \frac{2695}{\Delta l} \tag{2.9}$$

Thus, the range of spring stiffness values could be obtained as

$$\frac{2695}{\Delta l} \leq k \leq 4.3382 \times 10^4 \times f$$

The specific value needs to be determined in conjunction with the actual project. After processing, testing, and continuous improvement, currently designed isolator is shown in Fig. 2.20, which has four coils and eight sets of track plates, in which the wire diameter of the enameled wire on the winding plate is 1 mm. The total number of turns of the double-sided winding of the isolators is $N = 6720$, and the magnetic field strength has been measured as $B = 0.45T$ and $L = 0.1$ m, so that the electromagnetic force calculation formula is:

$$F = NBIL \tag{2.10}$$

where F is the electromagnetic force, N is the number of turns of the coil, B is the strength of magnetic field, I is the current, and L is the coil length in the magnetic field. It could be calculated that the electromagnetic force generated per unit current is 300 N.

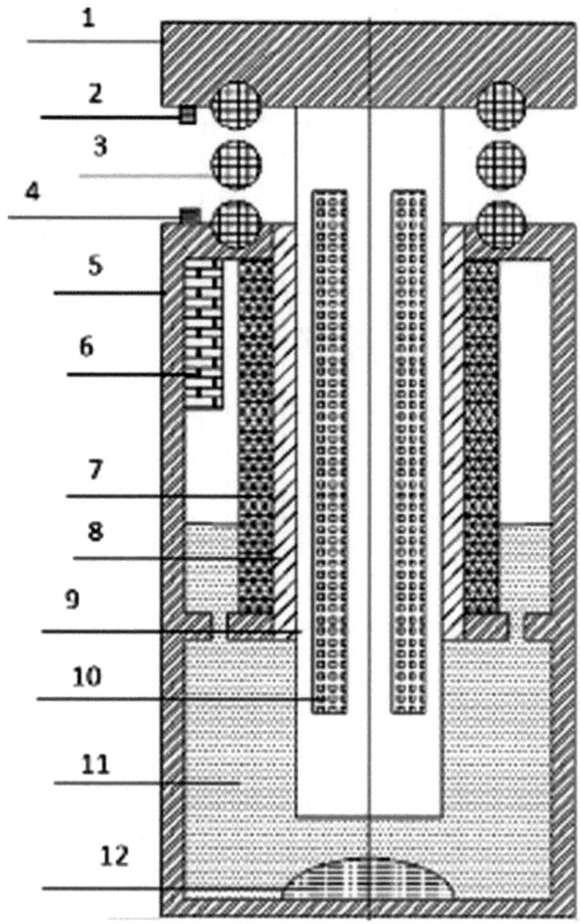

Fig. 2.18 Design schematic of active and passive hybrid actuators based on Halbach magnetic array

The prototype of the isolator is shown in Fig. 2.21, and the specific dimension is: 250 mm × 200 mm × 280 mm. The structure is an open design that fully considers the need for electromagnetically actuated heat dissipation. It facilitates the circulation and diffusion of heat. In addition, the arrangement of magnets and coils is parallel, which not only simplifies the processing and assembly, but also increases the stability and reliability of the structure and reduces the difficulty of post-maintenance. What is more, the slot design also facilitates the adjustment of electromagnetic force that depends on changes in operating conditions. The springs at the four corners can adjust the support height according to specific requirements, and an annularly arranged ball bearing is installed inside the guide column to reduce friction and power consumption.

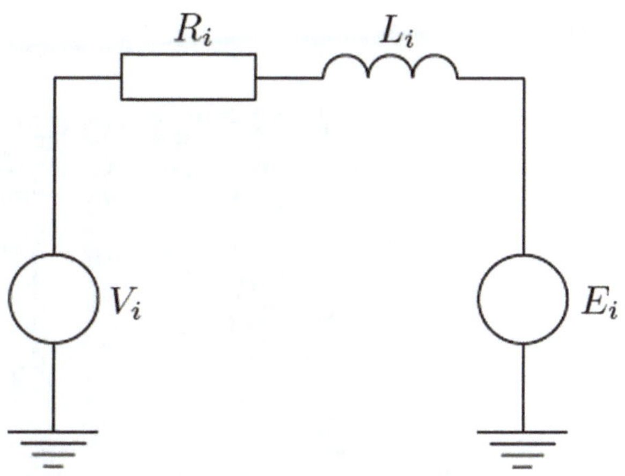

Fig. 2.19 Electric part of the circuit model

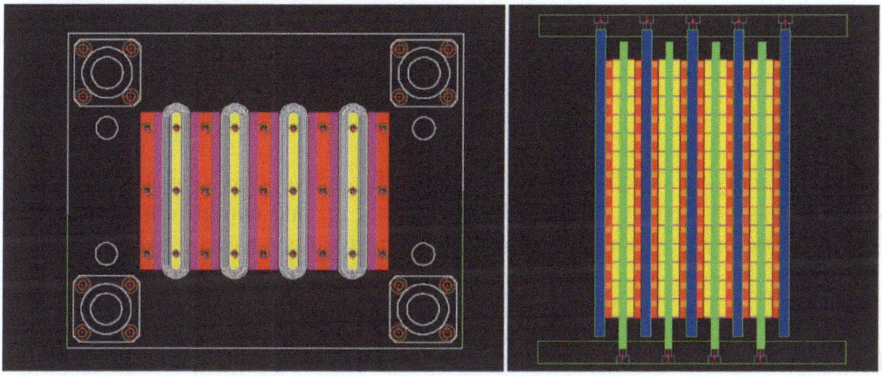

Fig. 2.20 Final design draft of active and passive hybrid vibration isolators based on Halbach magnetic array

2.5 Conclusion

In this chapter, the development status of passive vibration isolation, semi-active isolation, and active vibration isolation was briefly described. As a result, active vibration isolation is the most effective way to control low frequency vibration. What is more, to acquire a satisfactory vibration isolation effect in whole frequency band, active–passive vibration isolation is the primary choice.

Due to the excellent performance of Halbach magnetic array, it was used to design active–passive isolator. Firstly, the dynamic behavior of WD618 diesel engine was

Fig. 2.21 Prototype of active and passive hybrid isolators

tested and analyzed. Then performance index that active–passive isolator should be satisfied was derived. A design scheme was set up and carried out step by step. Finally, a prototype was manufactured successfully after several trials.

References

1. Dickens JD, Norwood CJ (1996) Vibration isolator facility. Department of Defence, Defence Science and Technology Oganisation
2. Goyder HGD, White RG (1980) Vibrational power flow from machines into built-up structures, part II: Wave propagation and power flow in beam-stiffened plates. J Sound Vib 68(1):77–96
3. Zhu HC, He L, Huo R et al (2002) Power flow analysis in designing the vibration isolation systems for marine main propulsion engines. In: National conference on vibration engineering and applications
4. Wu WP (2008) Study on active vibration control strategy of complex vibration isolation system. Shandong University
5. Xie S, Or S W, Chan H L W, et al (2007) Analysis of vibration power flow from a vibrating machinery to a floating elastic panel. Mech Syst Signal Process
6. Yan JK, Shen MQ (1982) How to use vibration isolator. Noise Vib Control 6:3–9
7. Yan JK, Shen RY (1986) Isolator selection and layout. Noise Vib Control 3:60–65
8. Zhou XR (2009) Research on the installation technology of setting damper. J Jiangsu Teach Univ (Natural Science Edition) 3:47–52
9. Shu LH, Hu ZC, Lv ZQ (2006) Overseas research progress on vibration isolator. Ship Sci Technol 28(3):109–112
10. Song YC, Yu HL (2007) Research of optimum selection scheme of isolator. J Dalian Marit Univ 33(1):87–89

11. Gu NC, Zhang YS (2007) Study on the arrangement and installation of vibration isolators. Ship Eng 29(2):30–33
12. Song WY, Li ZH (2002) The research of parameters choice of double-stage vibration arrester system. J Liaoning Tech Univ (Natural Science) 21(1):46–48
13. Sun YH, Dong DW, Yan B et al (2013) Study on finite element modeling methods of two-stage vibration isolation system. Mach Des Manuf 1:244–247
14. Ma YT, Zhou Y (2008) Summary of floating raft system. Ship Sci Technol 30(4):22–26
15. Wang ZX (1986) Spring design manual. Shanghai Science and Technology Literature Publishing House
16. Zhao G, Liu J, Liu ZS (2010) Theoretical and experimental study on nonlinear dynamic model of a rubber isolator. J Vib Shock 29(1):173–177
17. Wang R, Li SQ, Song SY (2006) Research on serialization design method of rubber vibration isolators. Noise Vib Control 26(4):11–13
18. Wu HL, Dai HJ (2009) Application of finite element analysis in design and development of rubber isolators. Noise Vib Control 29(1):114–116
19. Shi F, Tong ZP, Gong LQ et al (2009) Prediction of aging life for rubber vibration isolator. Ship Eng 31(4):42–44
20. Zheng XL (1983) Rubber isolator applications. Rubber Ind Des 2:35–41
21. Miao JM (2010) Design of rubber vibration isolator of electric equipment in warship. Mach Manag Dev 25(3):28
22. Li YY (2009) Application of metal rubber isolator. Fly Missile 5:62–63
23. Zhao SP, Liu FM (1994) New isolator and its application in environmental engineering. Environ Sci 1:65–68
24. Richards CM, Singh R (2001) Characterization of rubber isolator nonlinearities in the context of single-and multi-degree-of-freedom experimental systems. J Sound Vib 247(5):807–834
25. Kim BK, Youn SK, Lee WS (2004) A constitutive model and FEA of rubber under small oscillatory load superimposed on large static deformation. Arch Appl Mech 73(11):781–798
26. Lin CR, Lee YD (1998) Effects of viscoelasticity on rubber vibration isolator design. J Appl Phys 83(12):8027–8035
27. Chavan VS, Askhedkar R, Sanap SB (2013) Analysis of anti vibration mounts for vibration isolation in diesel engine generator set. Int J Eng Res Appl (IJERA), 1423–1429
28. Xu W, He L, Lv ZQ et al (2007) Analysis of dynamic characteristics of shipboard airbag vibration isolation system. J Vib Shock 26(7):122–124
29. Xiang F (2001) Block-oriented nonlinear control of pneumatic actuator systems. Maskinkonstruktion
30. Chen LC (1984) Experimental investigation of active pneumatic suspensions. Massachusetts Institute of Technology
31. Zhou T, Liu QL (2007) Simplified model analysis of wire-rope vibration isolator. J Vib Shock 26(9):55–59
32. Shu LH, Zhou W, Lv ZQ et al (2006) Stainless steel wire-rope isolator used in vibration and impact isolation design for large machine equipment. J Vib Shock 25(4):78–81
33. Tao X (2009) Research on property for wire-rope vibration isolation. Mach Manuf Autom 38(4):22–23
34. Liu GP, Wang FM, Fan WX (1999) Experimentalal study on the dynamic characteristics of steel cable isolators. J Test Meas Technol NCIT 13(3):180–184
35. Mizuno T, Toumiya T, Takasaki M (2007) Vibration isolation system using negative stiffness. JSME Int J 73(4):418–421
36. Sciulli D (1997) Dynamics and control for vibration isolation design. Virginia Tech. Dissertation
37. Giua A, Melas M, Seatzu C et al (2004) Design of a predictive semiactive suspension system. Veh Syst Dyn 41(4):277–300
38. Shan SJ, He L (2006) Study on controllable damping semi-active impact isolation technology. J Vib Shock 25(5):144–147

39. Maciejewski I, Krzyżyński T (2011) Control design of semi-active seat suspension systems. In: Proceedings of the 5th working IEEE/IFIP conference on software architecture. IEEE Computer Society, pp 261–262
40. Ahuja AS, Gupta A (2014) Fuzzy logic controlled semi-active floating raft vibration isolation system. Univ J Mech Eng 2(4):142–147
41. Bing Z et al (2015) The vibration isolation technologies of load in aviation and navigation. Int J Multimed Ubiquitous Eng 10(12):19–26
42. Jian-ye D, Yi Z, Hongxing H (2005) The application of mixed passive-active control technique to ship equipment vibration isolation and noise reduction: a review. In: Twelfth international congress on sound and vibration
43. Huang QF (2010) Analysis of the whole vibration process for AVS structural control system. J Huaqiao Univ (Natural Science) 31(1):83–87
44. Long ZQ, Hao AM, Chen G et al (2003) The research of active isolation platform with magnetically levitated control. J Astronaut 24(5):510–514
45. Hoque ME, Takasaki M, Ishino Y et al (2006) Development of a three-axis active vibration isolator using zero-power control. IEEE/ASME Trans Mechatron 11(4):462–470
46. Yang P, Liu F, Liu Y, et al (2008) Computer-aided design integration of a reinforced vibration isolator for electronic equipment's system based on experimental investigation. Struct Multidiscip Optim 35(5):489–498
47. He L, Li Y, Shuai C (2015) Active-passive vibration isolation for ship machinery using electromagnetic actuator and air spring. In: ICSV
48. Hoque ME, Takasaki M, Ishino Y et al (2006) An active micro vibration isolator with zero-power controlled magnetic suspension technology. JSME Int J Ser C Mech Syst Mach Elem Manuf 49(3):719–726
49. Lin H, McInroy JE (2003) Adaptive sinusoidal disturbance cancellation for precise pointing of Stewart platforms. IEEE Trans Control Syst Technol 11(2):267–272
50. El-Sinawi AH (2004) Active vibration isolation of a flexible structure mounted on a vibrating elastic base. J Sound Vib 271(1):323–337
51. Baig RU, Pugazhenthi S (2015) Design optimization of Stewart platform configuration for active vibration isolation. Indian J Sci Technol 8(23)
52. Liu YG, Zhang L, Fu YL et al (2004) A new adaptive feedforward active vibration isolation control technology. In: China Aviation Society annual conference on control and application
53. Chen B (2008) Floating isolation system modeling and active vibration control. University of Science and Technology of China
54. Li KQ (2008) Application of 6-RSS parallel mechanism in six-dimensional active vibration reduction platform. Beijing Jiaotong University
55. Yoshioka H, Murai N (2002) An active microvibration isolation system. J Vib Acoust 123(2):269–275
56. Muller T et al (2005) Modelling and control techniques of an active vibration isolation system. In: IMAC-XXIII
57. Singh R, Kim S (2003) Examination of multi-dimensional vibration isolation measures and their correlation to sound radiation over a broad frequency range. J Sound Vib 262(3):419–455
58. Aso Y (2008) Active vibration isolation for a laser interferometric gravitational wave detector using a suspension point interferometer. Ph.D. thesis, University of Tokyo
59. Arias-Montiel M, Silva-Navarro G, Antonio-García A (2014) Active vibration control in a rotor system by an active suspension with linear actuators. J Appl Res Technol 12(5):898–907

60. Ahn KK (2014) Active pneumatic vibration isolation system using negative stiffness structures for a vehicle seat. J Sound Vib 333(5):1245–1268
61. Jansson F, Johansson O (2003) A study of active engine mounts. Linköpings universitet, Sweden
62. Hassan Aly S (1968) Fundamental studies of passive, active and semi-active automotive suspension systems. University of Leeds
63. Widrow B, Hoffman M (1960) Adaptive switching circuits. In: Proceedings of the IRE WESCON convention record, Part 4, Session 16, pp 96–104
64. Beltran-Carbajal F, Silva-Navarro G, Abundis-Fong HF (2015) Application of passive/active duffing vibration absorbers in duffing mechanical systems. In: ICSV22

Chapter 3
Active and Passive Hybrid Vibration Isolator Performance Test

Abstract In this chapter, the following objectives are tested for the performance of the designed active–passive hybrid vibration isolator:

- Test spring stiffness, electromagnetic force to verify compliance with design requirements;
- Test whether insulation resistance meets the need for safe operation;
- Test resistance and inductance to design suitable power amplification equipment.

Furthermore, the active and passive vibration isolation performance of active-passive vibration isolator for single-frequency excitation will be tested.

3.1 Spring Stiffness Measurement

The actuator was fixed on the loader, and the load of 250–350 kg (mass of one quarter of the WD618 marine diesel engine) was loaded on the loader. The displacement of the upper surface was measured by a micrometer and converted to stiffness. The test setup is shown in Fig. 3.1.

When the pressure is changed from 2500 to 3500 N, the displacement and stiffness of the isolator roof are changed as is shown in Table 3.1 and Fig. 3.2. It could be seen that the stiffness of the spring is about 40 kgf/mm. Consider the frictional force existing in the structure under the condition of microdeformation. With the transformation between static friction and pressure changes, weak stiffness changes occur, but the relationship between stiffness and pressure is basically linear.

F. Wang et al., *Comprehensive Investigation on Active-Passive Hybrid Isolation and Tunable Dynamic Vibration Absorption*, Springer Tracts in Mechanical Engineering, https://doi.org/10.1007/978-981-13-3056-8_3

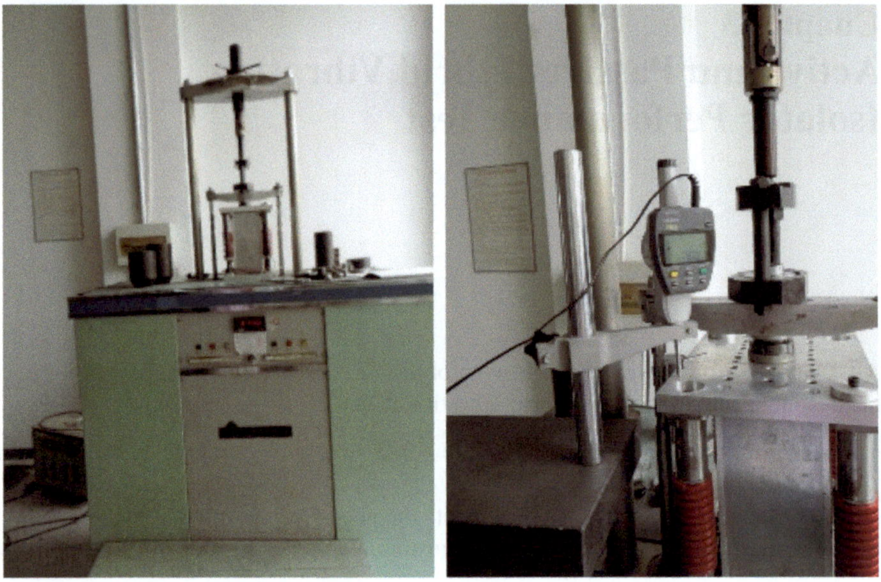

Fig. 3.1 Way to test the stiffness of isolator

Table 3.1 Test results of stiffness of isolator

Pressure (N)	Relative displacement (mm)	Stiffness (N/mm)
2500	0.000	452.49
2600	−0.221	393.70
2700	−0.475	408.16
2800	−0.720	381.68
2900	−0.982	384.62
3000	−1.242	420.17
3100	−1.480	495.05
3200	−1.682	375.94
3300	−1.948	416.67
3400	−2.188	373.13
3500	−2.456	–

3.2 Resistance and Inductance Measurements

3.2.1 Insulation Resistance

The requirements for insulation resistance are:

Insulation resistance of motor, power distribution equipment, and distribution line should not be lower than 0.5 at room temperature. If the insulation resistance is

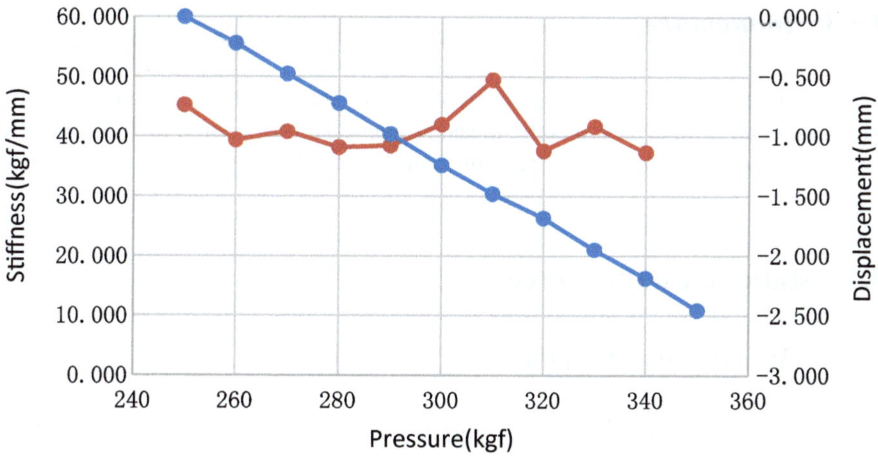

Fig. 3.2 Relationship between stiffness and displacement of isolator (1 kgf = 10 N)

Table 3.2 Test results of insulation resistance

Voltage (V)	Insulation resistance (MΩ)
200	12 ~ 40
500	17 ~ 50

Table 3.3 Test results of DC resistance

Coil 1#	4.334 Ω
Coil 2#	4.230 Ω
Coil 3#	4.139 Ω
Coil 4#	4.275 Ω

too low, it may cause energizing circuit with the surrounding equipment or ground during operation, thus affecting the surrounding equipment. The normal work may even cause damage to the equipment or endanger human safety in serious cases.

Using an insulation resistance tester, the insulation resistance between the isolator coil and the structure is measured, and the specific measurement results are listed in Table 3.2; it could be seen that the designed active and passive hybrid vibration isolator has a good insulation performance.

3.2.2 DC Resistance

Measure the DC resistance of the actuator coil with the ohm range of multimeter. The specific measurement results are listed in Table 3.3. It could be seen that the resistances of the four groups of coils are not much different, indicating good machining accuracy.

3.2.3 Inductance

Connect the four coils of the actuator in series, and measure the inductance with the RLC universal bridge. The specific measurement results are listed in Table 3.4. DC resistance and inductance measurements could be used to design power amplifiers.

3.3 Static Actuating Force

3.3.1 Measurement Method

The isolator is fixed on the platform, and the isolator is driven by a signal amplifier and a power amplifier. A pressure sensor is fixed on the top of the actuator, and the top of the pressure sensor is fixed with a pressure plate. The output signal of the pressure transmitter and receiver is conditioned by a charge amplifier and measured by a voltmeter. After the conversion, it is the static thrust of the isolator, as is shown in Fig. 3.3. Figure 3.4 shows the site layout.

3.3.2 Test Result

The isolators have four coils—2 and 2 in series, and the pre-tightening force between the actuator and the sensor is small. The sensitivity of the sensor is 2.72 pC/N, and the sensitivity of the charge amplifier is 1 mV/N. The specific test results are listed in Table 3.5.

Table 3.4 Test results of inductance of isolator

Test condition	Coil 1# (mH)	Coil 2# (mH)	Coil 3# (mH)	Coil 4# (mH)	Four coils in series (mH)	Two coils in series and two in parallel (mH)
Frequency 100 Hz/bridge voltage 1 V	19.36	18.29	18.67	18.60	74.87	18.83
Frequency 1 kHz/bridge voltage 1 V	3.04	2.96	3.05	2.98	11.85	2.96
Frequency 10 kHz/bridge voltage 1 V	2.23	2.20	2.29	2.22	9.20	2.21

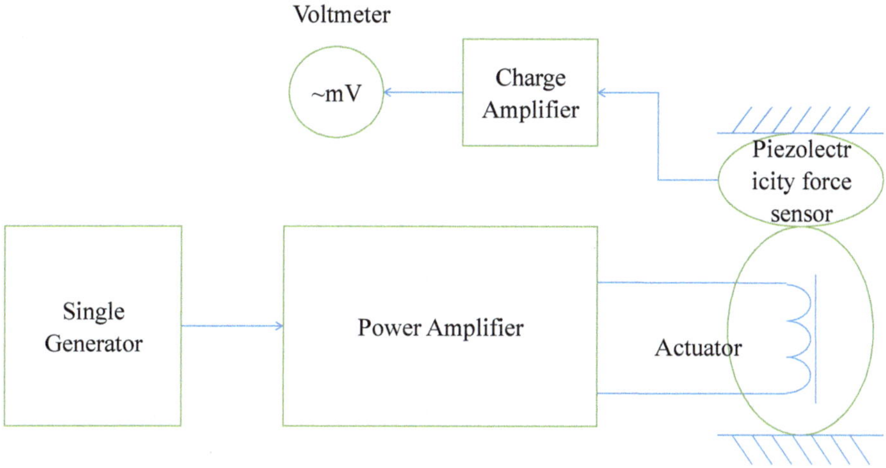

Fig. 3.3 Schematic of testing static thrust

Fig. 3.4 Site map of static thrust test of vibration isolator

Combining the above measurement results, the current–peak thrust curve of the isolator coil could be obtained as is shown in Fig. 3.5. Among them, before the current reaches 1.5 A, the measured working force is small, but this does not represent the actual electromagnetic force, because the measurement result is the resultant force of electromagnetic force and friction and gravity, that is,

$$F_{\text{Measure}} = F_{\text{act}} - F_{\text{friction}} - F_{\text{weight}} \tag{3.1}$$

Table 3.5 Test results of static thrust of isolator

Total voltage of four coils (V)	Current of single coil (A)	The output voltage of charge amplifier (mV)	Peak thrust (N)
10.18	0.505	3.7	5.2318
20.9	1.01	36.8	52.0352
30.06	1.5	175.3	247.8742
40.56	2.05	336	475.104
50.2	2.51	459	649.026
60.4	3.05	590	834.26
70.2	3.5	720	1018.08
80.2	4.01	870	1230.18
90.4	4.51	997	1409.758
100.04	5	1111.1	1571.0954

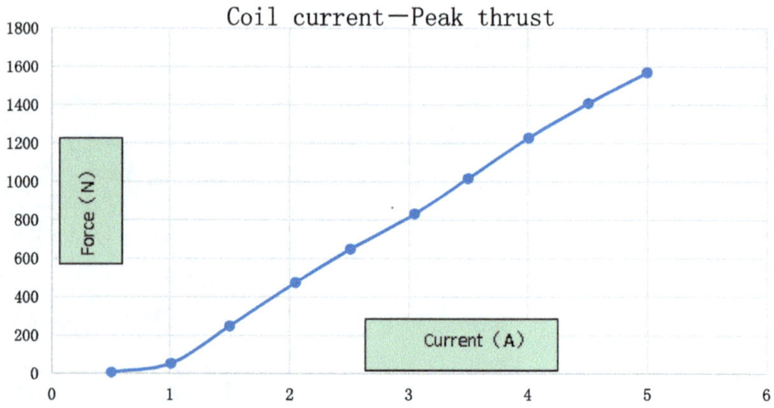

Fig. 3.5 Curve of isolator coil current–peak thrust

It could be seen from the figure that the dynamics of the isolator satisfy the design requirements, and the relationship between the power and the current could be further obtained as:

$$F_{\text{act}} = k_{\text{I}} \hat{i} + \text{d}v \approx 300 \hat{i} \tag{3.2}$$

where v is the speed of actuator. The electromagnetic force test results are consistent with the previous theoretical calculations.

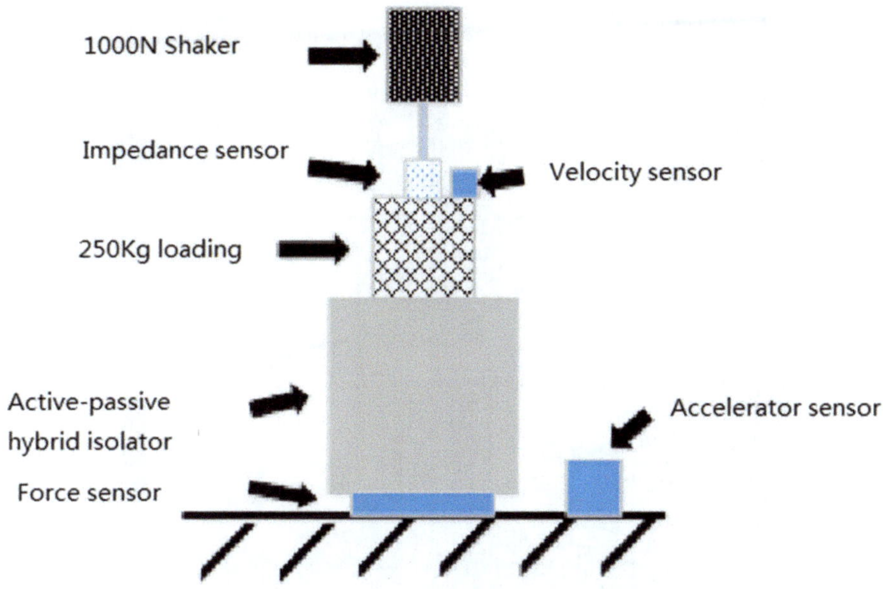

Fig. 3.6 Test schematic of single-frequency vibration isolation effect for active and passive hybrid vibration isolator

3.4 Vibration Isolation Performance Analysis

In front of the tests, the performance indicators of the isolator were verified. Results show that the designed isolator has excellent electromagnetic characteristics. This section will test the vibration isolation performance of a vibration isolator.

Using a 1000 N exciter as excitation, the impedance head (a sensor that can simultaneously measure force and velocity) is used to acquire the signal to achieve closed-loop control of the exciter, the accelerometer is used to measure the loaded vibration, and the 250 kg load is used to simulate a quarter-diesel engine. The mass, force plate, and acceleration sensors are used to calculate the force transfer rate and observe the basic vibration conditions, as is shown in Fig. 3.6.

The test site is shown in Fig. 3.7. Figure 3.8 shows the on-site installation and adjustment photographs during the test.

3.4.1 Passive Vibration Isolation Performance

The active–passive hybrid vibration isolator is used as a passive isolator, and active control is not turned on.

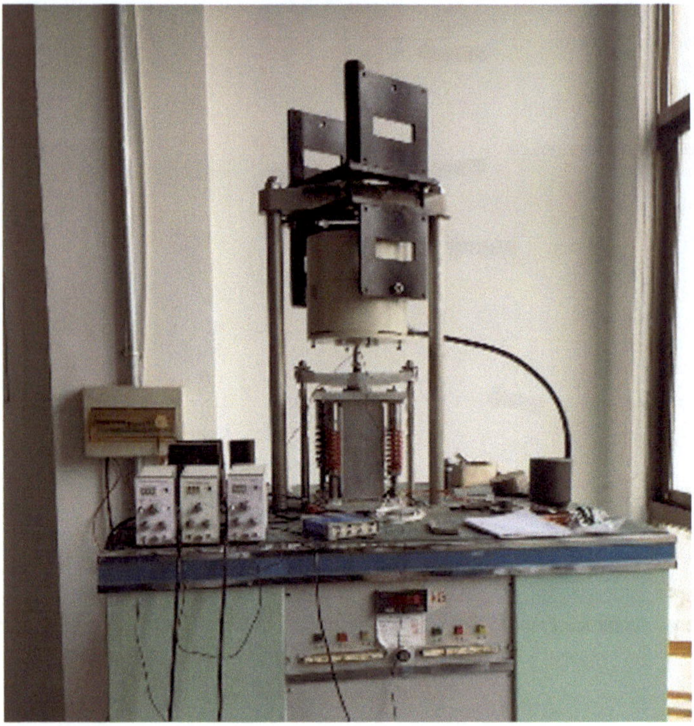

Fig. 3.7 Site map of test

A signal generator and a power amplifier connected to the exciter are set so that the exciter outputs a sine signal with a single frequency and a peak value of 600 N (the entire structure is destabilized at 1000 N).

After the system is stable, the response data of the impedance head, force plate, and acceleration sensor are collected (time domain signal, duration 60 s, repeated three times). Figures 3.9, 3.10, and 3.11 show the effect of vibration isolation when the excitation frequency is 25 Hz (the frequency corresponding to the rated diesel engine speed of 1500 r/min).

By analyzing the test signals, it could be found that the excitation power of the 600 of 25 Hz could be attenuated to about 1/2 of the original in the case of using only passive vibration isolation; that is, the vibration isolation effect is 6 dB.

Fig. 3.8 Local drawings of installation

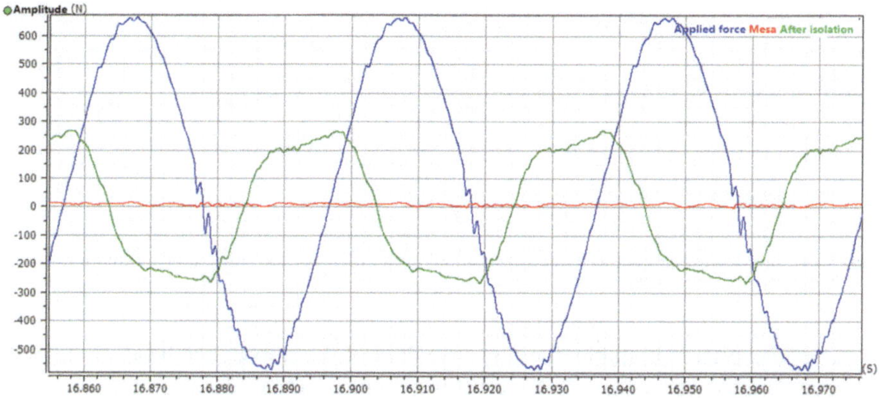

Fig. 3.9 Single-frequency passive vibration isolation effect for the first data acquisition results

3.4.2 Passive and Eddy Current Damping Vibration Isolation

Without active control, the positive and negative poles of the four groups of active and passive hybrid isolators are connected to form four closed loops to form an eddy current damper.

A signal generator and a power amplifier connected to the exciter are set so that the exciter outputs a sine signal having a single frequency and a peak value of 600 N.

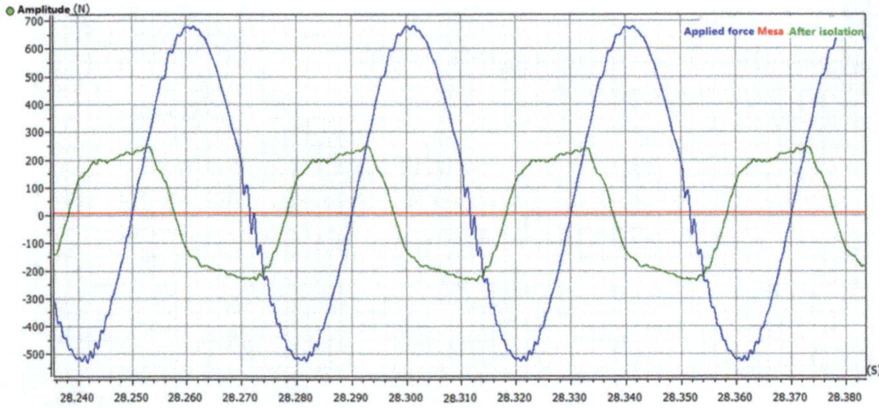

Fig. 3.10 Single-frequency passive vibration isolation effect for the second data acquisition results

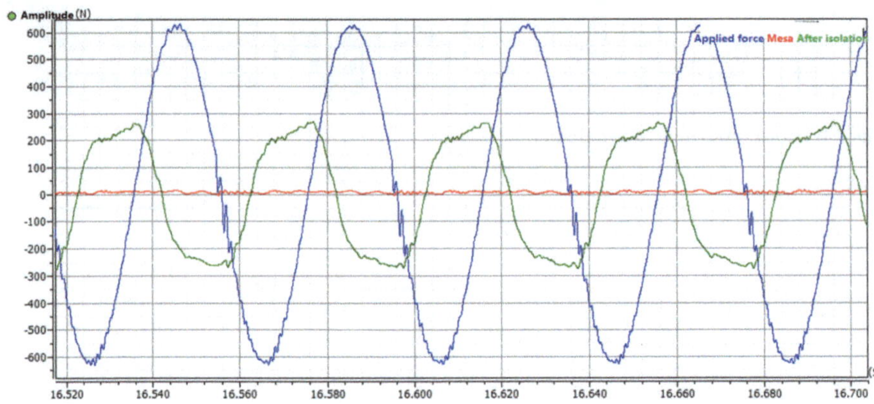

Fig. 3.11 Single-frequency passive vibration isolation effect for the third data acquisition results

After the system is stable, the response data of the impedance head, force plate, and acceleration sensor are collected (time domain signal, duration 60 s, repeated three times). Figures 3.12, 3.13, and 3.14 show the effect of vibration isolation at an excitation frequency of 25 Hz.

It could be found that after the damper damping function is activated, the vibration isolation effect is almost the same as the vibration isolation effect when the stiffness is only used for damping, because the amplitude of the isolator is too small and the eddy current damper needs a certain stroke to work. On the other hand, because the interference frequency is higher than the resonance frequency, the installation frequency currently is $f = \frac{1}{2\pi}\sqrt{\frac{K}{M}} = \frac{1}{2\pi}\sqrt{\frac{400000}{250+50}} = 5.81$ Hz and 250 and 50 units are kg, which, respectively, represent the quality of the loading and vibration isolator.

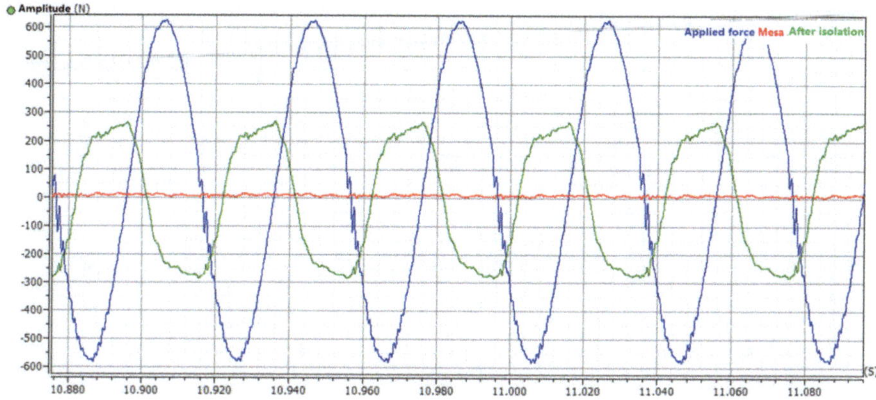

Fig. 3.12 Single-frequency passive vibration isolation and eddy current damping vibration isolation effect for the first data acquisition results

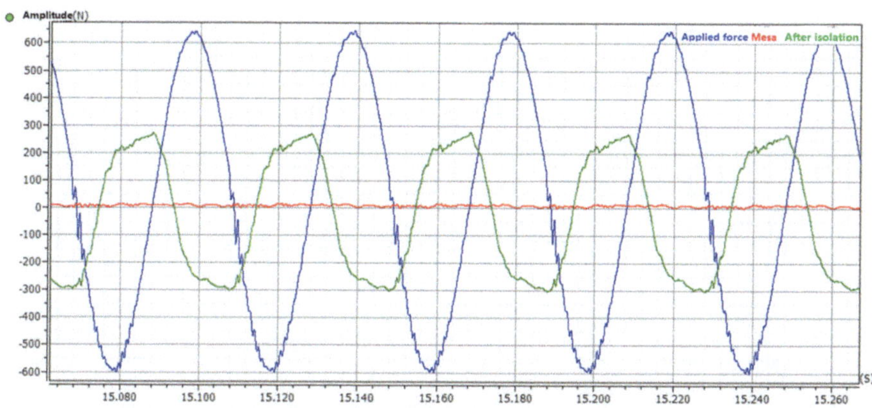

Fig. 3.13 Single-frequency passive vibration isolation and eddy current damping vibration isolation effect for the second data acquisition results

3.4.3 Active and Passive Hybrid Vibration Isolation

A signal generator and a power amplifier connected to the exciter are set so that the exciter outputs a sine signal with a single frequency and a peak value of 500 N.

Since the purpose of the test is only to verify the ability of the isolator to control the low-frequency line spectrum, it is not necessary to use complex control algorithms. The excitation signal of the isolator could be obtained by phase-shifting the excitation signal of the exciter, through the observation of the force plate. Outputting and properly adjusting the digital amplifier gain and the phase difference between the two excitation signals, you can get the best vibration isolation effect.

After the system is stable, the response data of the impedance head, force plate, and acceleration sensor are collected (time domain signal, duration 60 s, repeated three

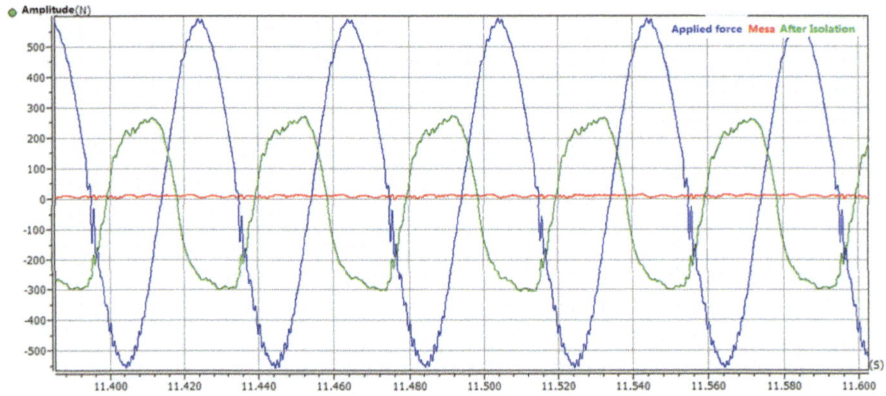

Fig. 3.14 Single-frequency passive vibration isolation and eddy current damping vibration isolation effect for the third data acquisition results

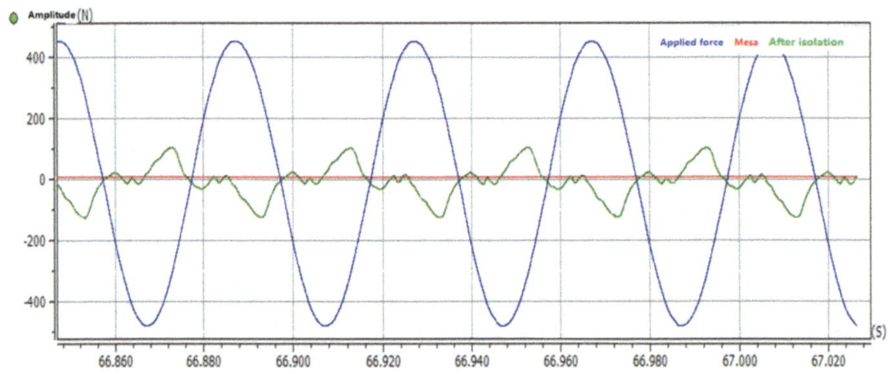

Fig. 3.15 Active and passive hybrid vibration isolation effect for the first data acquisition results

times). And collect data. Figures 3.15, 3.16, and 3.17 show the vibration isolation effect when the excitation frequency is 25 Hz.

Through observation, we can find that by adjusting the gain and the phase difference of the excitation signal, you can obtain a better damping effect; the active vibration isolator can attenuate the vibration of the exciter to 1/7; i.e., the isolation effect is 17 dB, obviously better than passive vibration isolation.

It should be noted that the active controller is not used in the test of the active–passive hybrid vibration isolation performance because the frequency and amplitude of the known interference signal are known, and the signal is single frequency, and the adjustment is controlled by observing the output signal of the force plate. The phase and amplitude of the signal generator of the actuator can achieve a better vibration isolation effect. This control method is an adaptive feed-forward control but uses the human brain instead of the controller function.

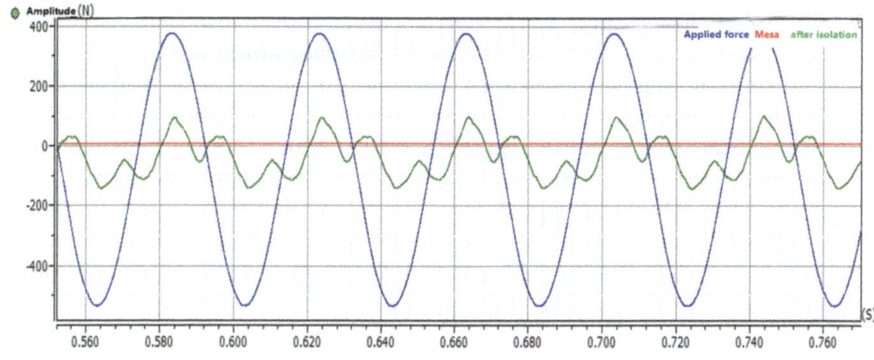

Fig. 3.16 Active and passive hybrid vibration isolation effect for the second data acquisition results

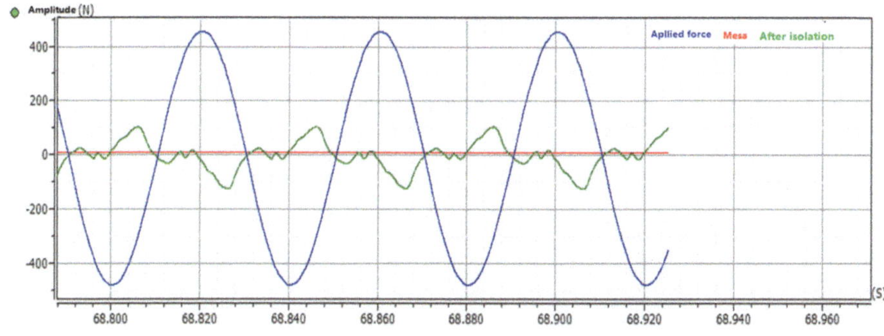

Fig. 3.17 Active and passive hybrid vibration isolation effect for the third data acquisition results

3.5 Conclusion

In this chapter, the active and passive hybrid vibration isolator based on the Halbach magnetic array design is proposed. The design method and scheme of the active and passive hybrid vibration isolator for the vibration characteristics of the diesel engine are presented. The active vibration isolation unit is processed and assembled, and its resistance, inductance, stiffness, and static thrust are tested. The curve of the relationship between the coil current and the peak thrust of the isolator was obtained, and the vibration isolation performance of the isolator was measured using the excitation method. Research shows:

(1) The active and passive hybrid vibration isolator based on the Halbach magnetic array design has a simple structure. Four sets of coils are arranged in parallel with the magnetic array, and each unit of current can output 300 N. The theoretical calculation is consistent with the test results. With open characteristics, it can greatly increase the heat dissipation area, reduce the difficulty of processing and assembly, and improve the stability, reliability, and maintainability of the structure.

(2) The isolator has not only good insulation properties and excellent stability, but also a design requirement of carrying 300 kg and a power of 1500 N. There is a good linear relationship between the input current and the output power.
(3) The single-frequency passive vibration isolation of the vibration isolator can attenuate the excitation force by 1/2 (6 dB). After the damping function is activated, the vibration isolation effect is slightly better than the vibration isolation effect when only the spring vibration is used. After the active vibration isolation is turned on, the single-frequency active–passive vibration isolation test can attenuate the vibration of the exciter to 1/7 (17 dB), which is obviously better than that of the passive vibration isolation.

Chapter 4
Adaptive Feed-Forward Control System

Abstract This chapter designs and implements the active and passive hybrid vibration isolation control system. The complete active control system includes software and hardware. The software mainly refers to the control strategy and control algorithm implemented in code form. The hardware includes sensors, actuators, power amplifiers, and controllers. This chapter is based on the vibration characteristics and vibration transmission control characteristics of the WD618 marine diesel engine. The hardware design of the active control system includes a controller and a digital power amplifier that satisfies the power requirements of the active and passive hybrid isolator.

4.1 Introduction

In Chaps. 2 and 3, aiming at the vibration characteristics of the diesel engine under typical operating conditions, the active and passive hybrid vibration isolator is designed based on the Halbach magnetic array. The single-frequency active and passive vibration isolation performance of the isolator is measured by using the excitation method. To realize the active and passive vibration isolation based on adaptive feed-forward control under the condition of simulating the actual use environment, an active control system needs to be established. The complete active control system includes software and hardware. The software system mainly refers to the control strategy and control algorithm implemented in the form of code. The hardware system includes sensors, power amplifiers, and controllers in addition to actuators. The control system collects the vibration signal and controls the actuator operation through a certain algorithm to reduce the vibration at the target position.

This chapter will study the adaptive feed-forward control algorithm based on the characteristics of active vibration isolation of diesel engines, propose an adaptive feed-forward control method combining the characteristics of active and passive hybrid vibration isolation, and use the measured data to simulate the control effect of LMS algorithm and RLS algorithm. The hardware systems such as power

F. Wang et al., *Comprehensive Investigation on Active-Passive Hybrid Isolation and Tunable Dynamic Vibration Absorption*, Springer Tracts in Mechanical Engineering, https://doi.org/10.1007/978-981-13-3056-8_4

amplifiers and controllers are designed to solve the problems of large, heavy, and heat-dissipating analog power amplifiers.

4.2 Feed-Forward Control for Active Vibration Isolation

For reciprocating or rotating equipment, feed-forward control is usually selected because a stable reference signal could be obtained. Feed-forward control can make full use of prior knowledge compared to feedback control, so the control effect for line spectrum is better. In addition, considering the change of working conditions, the algorithm should have the ability of adaptive adjustment, so the adaptive feed-forward control is selected.

The feed-forward control is applied to the active vibration isolation. The control system consists of a reference sensor, an error sensor, a controller, a power amplifier, and an actuator. The purpose of the control is to adaptively adjust the coefficients of the control filter based on the input of the reference sensor and the error sensor. The output of the actuator can counteract the effects of the primary operating force, which results in less vibration at the error sensor, as is shown in Fig. 4.1.

For the convenience of analysis, Fig. 4.1 is further abstracted as is shown in Fig. 4.2. It is composed of primary disturbance, mechanical system, and controller. The digital controller H consists of an estimate of the primary excitation signal x (in practice, the reference sensor output, i.e., the reference signal) signal drive. At the same time, the error signal e proportional to the response of the mechanical system is used in the control channel to assist in adjusting the response of the controller; that is, the filter coefficient of the controller is adjusted based on the error signal to minimize the error.

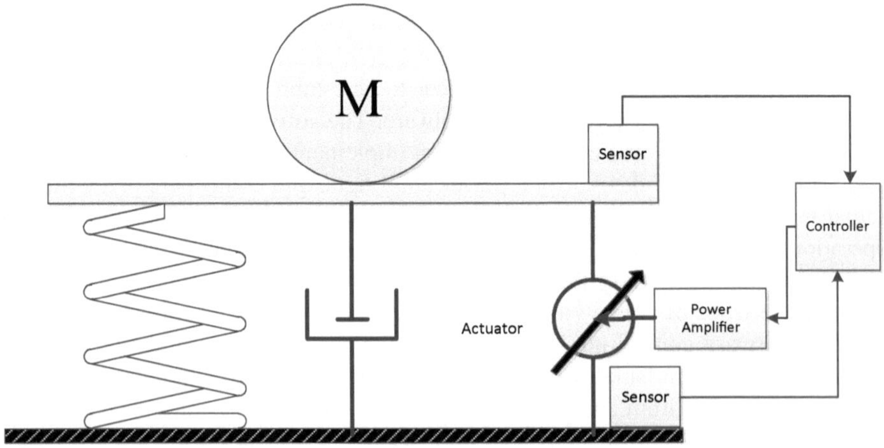

Fig. 4.1 Schematic of feed-forward control in active vibration isolation

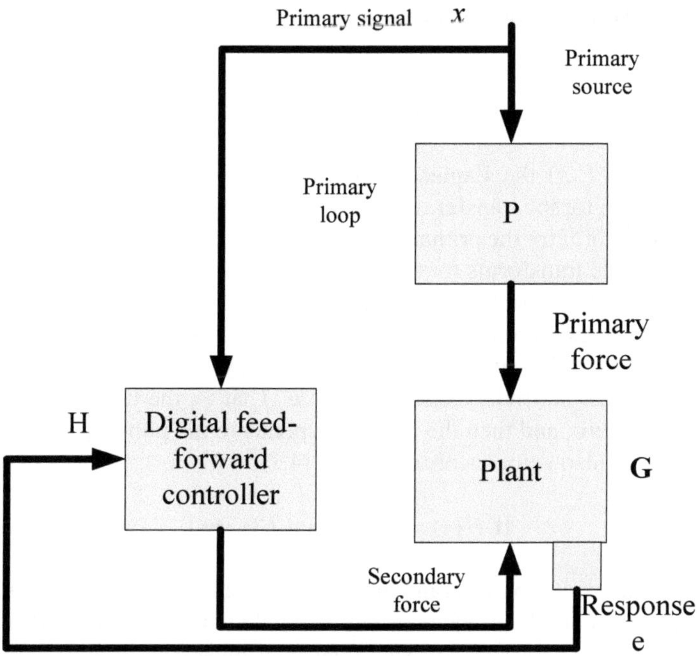

Fig. 4.2 Components of feed-forward control system

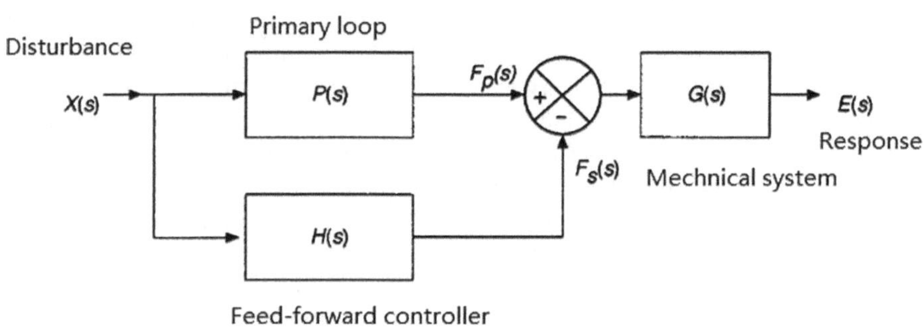

Fig. 4.3 Block diagram of feed-forward control system

Assuming that the primary force f_{p} causes disturbance to the system test through the primary channel, the net response of the mechanical system is the difference between the system's response to the primary disturbance force f_{p} and the secondary actuation force f_{s}.

Figure 4.3 is an equivalent block diagram of Fig. 4.2, where the signal represents the Laplace transform of the corresponding time domain signal, namely $P(s)$, $X(s)$, $H(s)$, $F_{\mathrm{P}}(s)$, $F_{\mathrm{S}}(s)$, $G(s)$, and $E(s)$ the Laplace transform quantities for P, x, H, f_{p}, f_{s}, G, and e, respectively.

According to Fig. 4.3, the Laplacian of the time domain response of the mechanical system is transformed to

$$E(s) = G(s)[P(s) - H(s)]X(s) \tag{4.1}$$

Among them, $E(s)$ the Laplace transform for the system response, $G(s)$ the Laplace transform for the transfer function of the mechanical system, $P(s)$ and $H(s)$ the Laplace transform for the primary channel and the digital controller, respectively, $X(s)$ is the Laplace transforms for the primary excitation. At this time, it is assumed that the response of the mechanical system is only caused by the initial excitation F_p and the secondary excitation F_s. In principle, if the influence of noise is neglected, the influence of the primary operating force on the system could be completely offset by adjusting the secondary operating force. That is, the time domain response of the system is zero, and then the system responds to the pull–pull. The amount of transformation is also zero, according to Eq. (4.1),

$$\text{If } H(s) = P(s), \text{ then } E(s) = 0 \tag{4.2}$$

If the original excitation is a random signal, then Eq. (4.2) must hold for all possible values of the complex frequency, indicating that the amplitude and phase of the frequency response of the feed-forward controller must be the same as the primary channel at all frequencies. In theory, this is only a problem with the design of electronic filters; however, many problems will be encountered in practice, especially when implementing feed-forward controllers in digital form, there will be unavoidable time lags. The lag causes the digital controller to not model the initial part of the impulse response of the primary channel (causality needs to be met).

However, for deterministic interference, causality problems are not particularly serious because theoretically the future characteristics of interference could be predicted. For example, for a sinusoidal interference with an angular frequency of sinusoidal ω_0, the reference signal is selected to be unit complex sine, i.e., $X(j\omega_0) = e^{j\omega_0 t}$, the complex response of the mechanical system could be represented as

$$E(j\omega_0) = G(j\omega_0)[P(j\omega_0) - H(j\omega_0)] \tag{4.3}$$

In order to accurately cancel this frequency response, the amplitude and phase of the controller at the frequency ω_0 need to be adjusted to be equal to the amplitude and phase of the primary channel, i.e., for the frequency ω_0,

$$\text{If } H(j\omega_0) = P(j\omega_0), \text{ then } E(j\omega_0) = 0 \tag{4.4}$$

Obviously, this condition is not difficult to achieve for single frequency ω_0. However, it should be noted that if it is necessary to achieve attenuation of 20 dB, then the amplitude response of the controller's complex response $H(j\omega)$ to the complex response of the primary channel $P(j\omega)$ must not exceed a ± 0.6 dB, and the phase deviation should not exceed $\pm 4°$.

4.3 Adaptive Filter

For the digital control system shown in Fig. 4.4, if the system is causal; that is, the output of the system is not leading the input of the system, and the output sequence $y(n)$ is only related to the current value and past value $x(n-1), \ldots$, i.e.,

$$y(n) = H[x(n), x(n-1), \cdots] \tag{4.5}$$

Among them, the digital controller is a digital filter H. If the digital system is linear the superposition theorem is satisfied, the linear sum of the sums could be used to express the function H; and for a linear causal system, the output signal is related to all past signals, i.e.,

$$y(n) = \sum_{i=0}^{\infty} h_i x(n-i) \tag{4.6}$$

That is, the output signal is a discrete-time convolution $x(n)$ with h_i.

The parameter h_i is the sampling of the system impulse response, i.e., the input sequence $x(n)$ is a Kronecker function,

$$x(n) = \begin{cases} 1 & n = 0 \\ 0 & n \neq 0 \end{cases} \tag{4.7}$$

Then,

$$y(n) = \sum_{i=0}^{\infty} h_i \delta(n-i) = h_n \quad n \in [0, \infty) \tag{4.8}$$

The stable system defined here means that the bounded input produces a bounded output (BIBO), which is a sufficient and necessary condition that the impulse

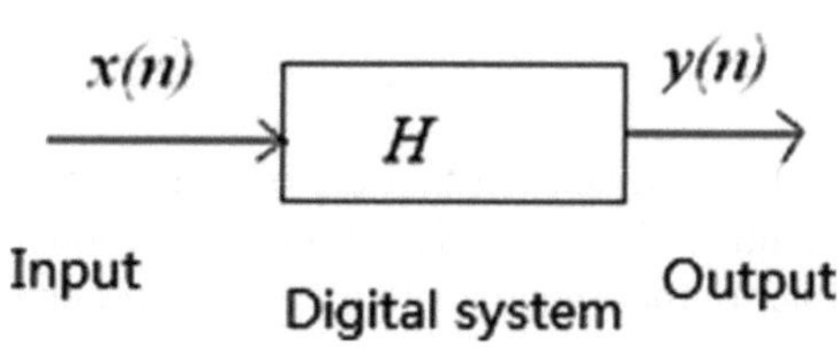

Fig. 4.4 General diagram of digital system

response sequence satisfies a complete additive sum, i.e., $x(n) = h_n$ when the following conditions are satisfied:

$$\sum_{n=-\infty}^{\infty} |x(n)| < \infty \tag{4.9}$$

In general, filters could be divided into FIR filters and IIR filters based on the duration of the filter response.

The FIR filter truncates (4.6) to

$$y(n) = \sum_{i=0}^{I-1} w_i x(n-i) \tag{4.10}$$

where w_i is the coefficient of the digital filter, also called the weight, assuming the order of this filter is I. Note that the output $y(n)$ currently depends on the current input $x(n)$.

The prerequisite for the establishment of (4.10) is that the digital filter can perform synchronous calculation and output, which is obviously impossible to achieve in the real-time cóntrol system. Therefore, it is generally assumed that there is a sampling delay in the real-time control system, which represents the processing time of the system, so that the output could be re-expressed as

$$y(n) = \sum_{i=0}^{I} w_i x(n-i) \tag{4.11}$$

Observation (4.11) shows that the response of the digital filter to the Kronecker impulse excitation $\delta(n)$ is a finite sequence, i.e., $y(n) = w_n, \quad 0 \leq n \leq I-1$ a filter of this type with finite impulse response, the so-called FIR filter.

Use z^{-1} to indicate unit delay, i.e.,

$$z^{-1}x(n) = x(n-1) \tag{4.12}$$

Expression (4.12) could be further expressed as

$$y(n) = W(z^{-1})x(n) \tag{4.13}$$

where

$$W(z^{-1}) = w_0 + w_1 z^{-1} + w_2 z^{-2} + \cdots + w_{I-1} z^{-I+1} \tag{4.14}$$

It could be seen that the output of the FIR filter is a weighted sum of a limited number of samples. The transfer function that relates the z transformation of the FIR filter output sequence to the z transformation of the input sequence is

$$Y(s) = W(s)X(s) \tag{4.15}$$

The simultaneous z-transform on both sides of (4.14) has

$$Y(z) = W(z)X(z) \tag{4.16}$$

where $Y(z)$ and $X(z)$ are the z transformations of sequences $y(n)$ and $x(n)$, respectively. The polynomial with (4.16) can also be expressed as a z polynomial $w_0 z^{I-1} + w_1 z^{I-2} + \cdots + w_{l-1}$, which is to be divided by z^{I-1}.

The FIR filter has the following important properties: (1) It is always stable when the coefficient is bounded; (2) small changes in the coefficient cause small changes in the response.

The general linear form of the IIR filter is

$$y(n) = \sum_{j=1}^{J} a_j y(n-j) + \sum_{i=0}^{I-1} b_i x(n-i) \tag{4.17}$$

Among them, there are J feedback coefficients a_j and I feed-forward coefficients b_i.

The IIR filter requires an infinite amount of time to attenuate the response to the pulse excitation; that is, it has an infinite impulse response.

The z conversion of (4.17) is

$$A(z)Y(z) = B(z)X(z) \tag{4.18}$$

where

$$A(z) = 1 - a_1 z^{-1} - \cdots - a_J z^{-J} \tag{4.19}$$

$$B(z) = b_0 + b_1 z^{-1} + \cdots + b_{I-1} z^{-I+1} \tag{4.20}$$

You can further organize the formula into

$$H(z) = \frac{Y(z)}{X(z)} = \frac{B(z)}{A(z)} \tag{4.21}$$

The above equation is the transfer function of the system defined by Eq. (4.18).

When using an FIR or IIR filter to accurately characterize the sampling of the system, the number of weights depends primarily on the nature of the desired physical system. If the system has a small amount of under-damped mode, the number of which has the same resonance frequency in response to the use of the IIR filter can usefully be characterized by the system characteristics. If a system has many over damped modes, no peaks appear in the frequency response, and a FIR filter could be used to better describe such systems.

Combining with the vibration characteristics of the diesel engine, it could be seen that the IIR filter is suitable for use, but the digital implementation of the IIR filter requires a long calculation time, so it is necessary to select according to the actual situation of the control system.

To minimize the vibration at the error sensor, the filter coefficients need to be adjusted to optimally match the mechanical system. The so-called optimal filter refers to a filter that can give optimal performance under a given condition. The optimal performance is usually defined as the mean square or H_2 norm of the error signal because this operation minimizes the effect of errors.

The error signal $e(n)$ is defined as the difference between the desired signal $d(n)$ and the reference signal x_i filtered by the w_i weighted FIR filter, i.e.,

$$e(n) = d(n) - \sum_{i=0}^{I-1} w_i x(n-i) \tag{4.22}$$

For convenience, the sum of $w_i x(n-i)$ in Eq. (4.22) could be expressed as the product of the vectors within the vector, i.e.,

$$e(n) = d(n) - \mathbf{w}^{\mathrm{T}}\mathbf{x}(n) = d(n) - \mathbf{x}^{\mathrm{T}}(n)\mathbf{w} \tag{4.23}$$

where

$$\mathbf{w} = \left[w_0 \; w_1 \cdots w_{I-1}\right]^{\mathrm{T}} \tag{4.24}$$

$$\mathbf{x}(n) = [x(n) \; x(n-1) \cdots x(n-I+1)]^{\mathrm{T}} \tag{4.25}$$

The performance function is defined as the mean square sum of the errors, i.e.,

$$J = E\left[e^2(n)\right] \tag{4.26}$$

Among them, E is expected operator. The goal is to find the filter coefficients $w_0, \cdots, w_{i-1}$ that minimize the values J. If $x(n)$ and $d(n)$ both are not fixed, then the weights of the filters are also functions of time. Here, for the convenience of analysis, it is assumed that all signals are fixed and each state traverses (when the sampling value of an arbitrary moment is the same as the value of a sample function along the time axis), the expectation is time-invariant, i.e., we can use average arithmetic to get. Therefore, the performance function defined by Eq. (4.26) is equal to the mean square value of the error signal.

Combined with (4.23), the performance function could be re-expressed as

$$J = \mathbf{w}^{\mathrm{T}}\mathbf{A}\mathbf{w} - 2\mathbf{w}^{\mathrm{T}}b + c \tag{4.27}$$

where

$$\mathbf{A} = E\left[\mathbf{x}(n)\mathbf{x}^{\mathrm{T}}(n)\right] \tag{4.28}$$

$$\mathbf{b} = E[\mathbf{x}(n)\mathbf{d}(n)] \tag{4.29}$$

$$c = E[\mathbf{d}(n)] \tag{4.30}$$

In Eq. (4.28), the matrix **A** is usually called the Hessian matrix, and its elements are the autocorrelation function values of the reference signal.

$$\mathbf{A} = \begin{bmatrix} R_{xx}(0) & R_{xx}(1) \cdots & R_{xx}(I-1) \\ R_{xx}(1) & R_{xx}(1) & \\ \vdots & \ddots & \\ R_{xx}(I-1) & & R_{xx}(0) \end{bmatrix} \tag{4.31}$$

where $R_{xx}(m)$ is the symmetric autocorrelation function of $x(n)$, defined in the entire real-time series as

$$R_{xx}(m) = E[x(n) + x(n+m)] = R_{xx}(-m) \tag{4.32}$$

The more general form of the performance function will contain an item proportional to the square of the filter weight $\omega^{\mathrm{T}}\omega$, i.e.,

$$J = E\left[e^2(n)\right] + \beta \mathbf{w}^{\mathrm{T}}\mathbf{w} \tag{4.33}$$

where β is a positive real number representing the coefficient of the weight value. Equation (4.23) can also be expressed as Eq. (4.27), when the Hessian matrix becomes

$$\mathbf{A} = \mathbf{R} + \beta \mathbf{I} \tag{4.34}$$

Among them, **R** is the autocorrelation matrix shown on the right side of Eq. (4.31), **I** is a unit matrix.

The elements **b** of the vector in (4.27) are the values of the cross-correlation function of the reference signal and the desired signal, i.e.,

$$\mathbf{b} = [R_{xd}(0), R_{xd}(1), \ldots, R_{xd}(I-1)]^{\mathrm{T}} \tag{4.35}$$

In the entire real domain, the stable time series is

$$R_{xd}(m) = E[x(n)d(n+m)]E[x(n-m)d(n)] \tag{4.36}$$

Finally, c is a real domain scalar constant whose value is equal to the mean square of the desired signal. Finally, c is a real domain scalar constant whose value is equal to the mean square of the desired signal.

When the performance function is expressed in the form of Eq. (4.27), the mean squared error is a quadratic function of the FIR filter weight. This quadratic function always has a minimum value and does not necessarily have a maximum value;

because when one of the filter coefficients becomes large or small, J becomes very large. In Eq. (4.27), it is assumed that the matrix **A** is positive definite, and then J has a unique minimum value. If **A** was given as (4.28), **A** could be positive definite (also called non-singular) or semi-positive definite, depending on the spectral density of the reference signal and the number of FIR filter weights. If the number of spectral components has at least half the number of filter weights, the reference signal is said to be continuously excited or "spectrum abundant", and the autocorrelation matrix given by (4.28) could be guaranteed to be positive definite, and thus (4.27) has a unique minimum value.

It is possible to obtain the filter coefficients when the mean squared error signal is reduced to a minimum by solving the partial derivative of the corresponding coefficient with the performance function and making the result equal to zero. Expressed in vector form, there is

$$\frac{\partial J}{\partial \mathbf{w}} = \left[\frac{\partial J}{\partial w_0}\frac{\partial J}{\partial w_1}\cdots\frac{\partial J}{\partial w_{I-1}}\right]^{\mathrm{T}} \tag{4.37}$$

Combined with the definition of the performance function J, formula (4.27), formula (4.37) could be further expressed as

$$\frac{\partial J}{\partial \mathbf{w}} = 2[\mathbf{A}\mathbf{w} - \mathbf{b}] \tag{4.38}$$

If the signal $x(n)$ is continuous, **A** satisfies non-singularity. The coefficient of the optimal filter could be obtained by zeroing each element in (4.38).

$$\mathbf{w}_{\text{opt}} = \mathbf{A}^{-1}\mathbf{b} \tag{4.39}$$

This type of filter, which has an optimal filter coefficient, is generally called a Wiener filter.

Using the definitions **A** and **b** of autocorrelation and cross-correlation functions, we can re-evaluate (4.39) as

$$\sum_{i=0}^{I-1} w_{i,\text{opt}} R_{xx}(k-i) - R_{xd}(k) = 0, \quad \text{for } 0 \le k \le I-1 \tag{4.40}$$

Equation (4.40) represents a discrete form of the Wiener–Hopf equation.

The cross-correlation vector representing the past value of the I reference signal and the error signal I could be further represented by Eq. (4.39) as

$$E[x(n)e(n)] = E\left[x(n)\left(d(n) - x^{\mathrm{T}}(n)w\right)\right] = \mathbf{b} - \mathbf{A}w \tag{4.41}$$

By adjusting the coefficients of the FIR filter so that it satisfies Eq. (4.40), all the elements represented by Eq. (4.41) are zero. Minimizing the mean squared error, the Wiener filter zeroes the cross-correlation function between the reference signal and the error signal on a scale equal to the length of the filter coefficients; thus, the residual error signal will no longer contain current and past $I - 1$ the reference signal, this is the so-called orthogonally principle.

The values of the autocorrelation matrix and the cross-correlation function can usually be obtained by estimating the measurement data, and then the elements **A** and **b** of the neutralization Eqs. (4.28) and (4.29) could be determined. The first calculated self-power spectral density $x(n)$ and the cross-spectrum between $x(n)$ and $d(n)$ the sums density, then Fourier transform could be used to get the correlation function. Through the average characteristics of these reference signals and expected signals, the coefficients of the Wiener filter could be calculated by combining Eqs. (4.31), (4.35), and (4.39). By substituting Eq. (4.39) into Eq. (4.27), the minimum value of the mean squared error value could be obtained directly.

$$J_{\min} = c - \mathbf{b}^{\mathrm{T}}\mathbf{A}^{-1}\mathbf{b} \tag{4.42}$$

Therefore, the residual mean squared error could be directly calculated by using the statistical characteristics of the reference signal and the desired signal. This could be very useful at the initial stage of the design because it gives the theoretically optimal value that could be obtained.

4.4 The LMS Algorithm

4.4.1 Basics

In practice, the autocorrelation function and cross-correlation function must generally be obtained by estimating the history signal of the reference signal and the desired signal. Moreover, the calculation of the optimal filter with I coefficients includes the $I \times I$ calculation of the inverses of the autocorrelation matrices. Even though this matrix has very special properties (Symmetric and Toeplitz), there are also effective algorithms that could be used to invert, but the calculation time is still proportional to the I^2 computational task. Moreover, if the matrix is ill, results obtained may be unstable.

Another way to calculate the filter coefficients is to use the data in turn to adjust the filter coefficients so that they evolve toward the direction of the smallest mean squared error. All filter coefficients for each new data set need to be adjusted. Compared to the calculation of the full data length used to calculate the true optimal filter coefficients, the amount of data required for each adjustment will be much smaller. It is a so-called adaptive filter. For a stable signal, it not only converges to an optimal filter, but it also converges to an optimal filter even if the autocorrelation property changes with the signal. Thus, for cases where the signal change rate is slower than the convergence rate, an adaptive filter could be used to track well.

The steepest descent method widely used in adaptive FIR filters is based on the quadratic nature of the error surface of the filter. The principle is that if the coefficients of the filter are adjusted by an amount proportional to the negative local gradient of the performance function, the coefficients must evolve in the direction of the global minimum. If the steepest descent method is used to adjust all the coefficients at the same time, the adjustment filter could be adjusted. The adaptive coefficient of the coefficient is expressed as

$$\mathbf{w}(\text{new}) = \mathbf{w}(\text{old}) - \mu\frac{\partial J}{\partial \mathbf{w}}(\text{old}) \tag{4.43}$$

where μ is the convergence factor.

With Eqs. (4.28) and (4.29), Eq. (4.38) could be re-expressed as

$$\frac{\partial J}{\partial \mathbf{w}} = 2E\left[\mathbf{x}(n)\mathbf{x}^{\mathrm{T}}(n)\mathbf{w} - \mathbf{x}(n)d(n)\right] \tag{4.44}$$

The error signal is

$$e(n) = d(n) - \mathbf{x}(n)^{T}\mathbf{w} \tag{4.45}$$

Thus, formula (4.44) could be re-expressed as

$$\frac{\partial J}{\partial \mathbf{w}} = -2E[\mathbf{x}(n)e(n)] \tag{4.46}$$

To truly achieve the steepest descent method, the expected value of the product of the error signal and the delayed reference signal needs to be estimated to obtain Eq. (4.44).

In addition, van der Sande [1] proposed that instead of using gradient average estimation to intermittently update the filter coefficients, it is better to use the gradient's instantaneous estimate to update the gradient (the so-called statistical gradient) at the sampling instant, and this update amount is equal to the instantaneous error. The derivative of the filter coefficient, i.e.,

$$\frac{\partial e^{2}(n)}{\partial \mathbf{w}} = -2\mathbf{x}(n)e(n) \tag{4.47}$$

Thus, the adaptive algorithm becomes

$$\mathbf{w}(n+1) = \mathbf{w}(n) + \alpha e(n)\mathbf{x}(n) \tag{4.48}$$

where $\alpha = 2\mu$ is the convergence coefficient. The general value meets the following conditions to ensure the convergence of the algorithm

$$0 < \alpha < 2/\lambda\text{max} \tag{4.49}$$

where λ_{max} is the largest eigenvalue of $\overline{x^2}$, which equals to $E[x^2(n)]$, i.e., the mean square value of $x(n)$, in practice, can generally be obtained by averaging the I past data points, which is the famous LMS algorithm. LMS algorithm is easy to implement, and it has numerical robustness. It is widely used in various fields.

4.4.2 Simulation and Analysis

For the diesel engine active vibration isolation system, design suitable controller needs to carry on the simulation analysis to the adaptive feed-forward control algorithm, on the one hand chooses the more suitable control algorithm from the LMS algorithm and the RLS algorithm, on the other hand determines through the simulation computation optimum controller design parameters for active vibration isolation of diesel engines.

Using MATLAB as a simulation platform, using the measured diesel vibration data, the LMS algorithm and the RLS algorithm were simulated separately.

Since the control frequency band of interest is within 200 Hz, it is necessary to first perform low-pass filtering operations on the reference signal and the desired signal. In addition, according to the feed-forward control theory, it is necessary to determine the reference signal and the desired signal. Here, the vertical vibration of the diesel engine is used as the reference signal, the vibration signal of the machine foot is used as the desired signal, and the rated operating condition (1500 RPM) is used as the simulation analysis condition.

Based on the LMS function in MATLAB, using the filtered desired signal and the reference signal, the control output and the residual error could be obtained, as is shown in Figs. 4.5 and 4.6, respectively, for the desired signal and the LMS-controlled control output. The residual error is comparing the time domain with the frequency domain, and it could be found that the residual error after the control becomes very small. For example, for 115 and 150 Hz, the amplitude of the controlled vibration is smaller than the original 1%; that is, the control effect is higher than 40 dB.

For the LMS algorithm, both the filter order and the update step length influence the control effect. Figure 4.7 shows the influence of the filter order on the error when $\mu = 0.0002$. It could be seen from the figure that the filter order is further increased and the system will become unstable; that is, the order of the filter is not as good as possible.

Figure 4.8 shows the effect of step size on the error when the filter order $N = 256$. It could be seen from the figure that reducing the step size does not necessarily reduce the error, while increasing the step size to a certain degree will make the system unstable.

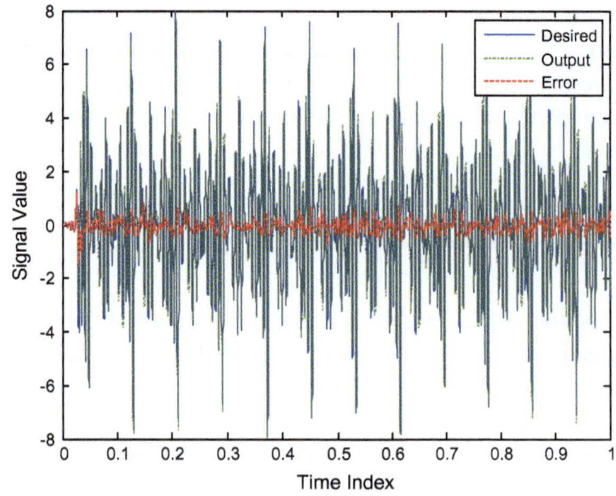

Fig. 4.5 Desired signal, LMS control signal, and error signal ($\mu = 0.0002$, $N = 256$)

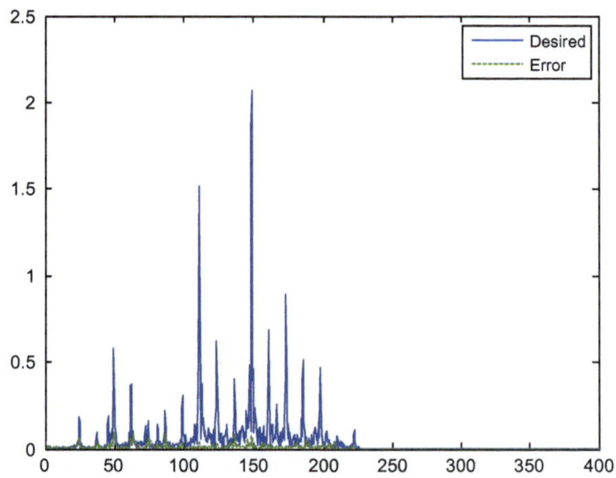

Fig. 4.6 Spectral comparison between desired signal and error signal ($\mu = 0.0002$, $N = 256$)

4.5 The RLS Algorithm

4.5.1 Basics

The slow convergence problem of the mode corresponding to the small eigenvalue of the Hessian matrix **A** is an inherent property of the steepest descent method. To avoid this problem, the filter coefficient could be updated using Newton's method, i.e., Eq. (4.43) could be expressed as:

$$\mathbf{w}(\text{new}) = \mathbf{w}(\text{old}) - \mu \mathbf{A}^{-1} \frac{\partial J}{\partial \mathbf{w}}(\text{old}) \tag{4.50}$$

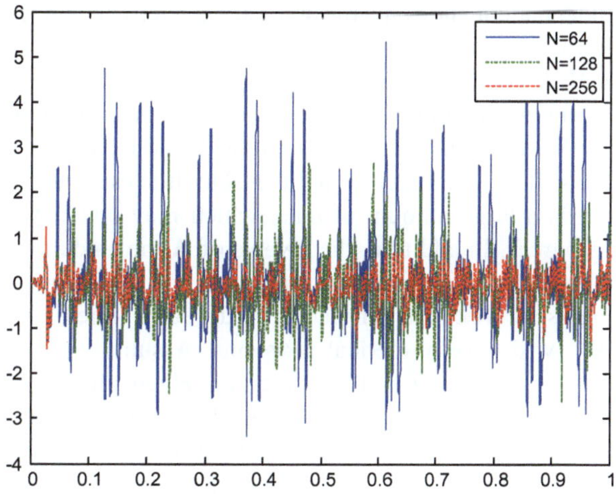

Fig. 4.7 Effect that filter order made on error ($\mu = 0.0002$)

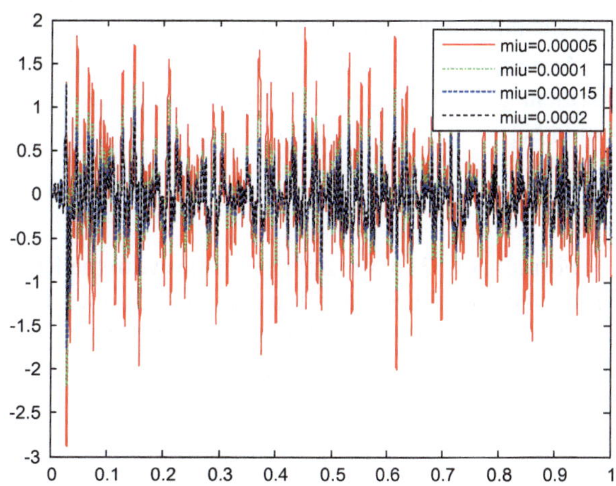

Fig. 4.8 Effect that step size made on error ($N = 256$)

Observing the above formula, the convergence characteristics of this algorithm could be obtained.

$$\frac{\partial J}{\partial \mathbf{w}} = 2[\mathbf{A}\mathbf{w} - \mathbf{b}] \tag{4.51}$$

Substituting this into Eq. (4.50), and noting that $\mathbf{A}^{-1}\mathbf{b} = \mathbf{w}_{\text{opt}}$, Eq. (4.50) could be reformulated as

$$\mathbf{w}(\text{new}) = (1 - \alpha)\mathbf{w}(\text{old}) + \mathbf{w}_{\text{opt}} \tag{4.52}$$

where $\alpha = 2\mu$. If we know the exact value of $\mathbf{A}^{-1}$ and combine the definitions $\partial J/\partial w$ at the same time, then using Newton's algorithm can make all the filter coefficients independent and converge at the same rate. The convergence of the small eigenvalues in the steepest descent method could be solved by using a predetermined gradient $\mathbf{A}^{-1}$. If you can accurately estimate the value $\partial J/\partial w$ in Eq. (4.51), you can achieve a single-step convergence to the optimal value by making μ equal to $\frac{1}{2}$.

However, in practice, estimating **A** and its inverse usually encounters many problems. If the reference signal is stable, it is estimated that the autocorrelation matrix needs a large amount of data, which may consume a large amount of computing resources. In addition, the inverse of **A** the calculation needs to be calculated. This calculation also consumes a large amount of computation time when the filter coefficients and both are very large. At the same time, if **A** is ill-conditioned, the solution $\mathbf{A}^{-1}$ will have numerical problems. Assuming that the eigenvalue of **A** equals to $\lambda_i, i = 0, 1 \ldots, I - 1$, and then the eigenvalue of $\mathbf{A}^{-1}$ is equal to $1/\lambda_i$; therefore, if some of the eigenvalues of **A** are extremely small, the calculated eigenvalue of $\mathbf{A}^{-1}$ will be wrong. Even to a certain extent, these problems could be solved by pre-calculating at the initial stage and using the fixed matrix in Eq. (4.50); however, this algorithm cannot adapt to the situation where the statistical properties change significantly; that is, it does not have real adaptive capabilities.

Assuming $\widehat{\mathbf{A}}^{-1}$ an estimate of $\mathbf{A}^{-1}$, and using the instantaneous estimate of the gradient given in Eq. (4.47) at each sampling instant, a modified Newton method could be

$$\mathbf{w}(n+1) = \mathbf{w}(n) - \alpha \widehat{\mathbf{A}}^{-1} \mathbf{x}(n) e(n) \tag{4.53}$$

Formula (4.53) has many similarities with the RLS—recursive least squares algorithm. For the RLS algorithm, the performance function reaches the minimum at each sampling instant. At the sampling time n, the performance function can usually be expressed as an exponentially weighted form of the mean squared error

$$J(n) = \sum_{l=0}^{n} \lambda^{n-l} e^2(l/n) \tag{4.54}$$

where $\mathbf{e}(l/n)$ is the variation of the error over time. When the filter coefficient is fixed and the current value, there is

$$\mathbf{e}(l/n) = d(l) - \mathbf{w}^{\mathrm{T}}(n)\mathbf{x}(l) \tag{4.55}$$

Note that the definition of $J(n)$ uses all past values of the variance $\mathbf{e}^2(l/n)$, gradually weighted to higher superscripts by the forgetting factor λ (between 0 and 1 and not confusing with the aforementioned eigenvalues).

The time-dependent performance function defined by Eq. (4.54) could be expressed in quadratic form, i.e.,

$$J(n) = \mathbf{w}^{\mathrm{T}}(n)\mathbf{A}(n)\mathbf{w}(n) - 2\mathbf{w}^{\mathrm{T}}(n)\mathbf{b}(n) + c(n) \tag{4.56}$$

where

$$\mathbf{A}(n)=\sum_{i=0}^{n} \lambda^{n-l}\mathbf{x}(l)\mathbf{x}^{\mathrm{T}}(l) \tag{4.57}$$

$$\mathbf{b}(n)=\sum_{i=0}^{n} \lambda^{n-l}\mathbf{x}(l)\mathbf{d}(l) \tag{4.58}$$

and

$$c(n)=\sum_{l=0}^{n} \lambda^{n-l}\mathbf{d}^{2}(l) \tag{4.59}$$

If the filter coefficient satisfies the following condition, Eq. (4.56) takes the minimum value at the sampling point; that is,

$$\mathbf{w}_{\mathrm{opt}}=\mathbf{A}^{-1}(n)\mathbf{b}(n) \tag{4.60}$$

It should be noted that $\mathbf{A}(n)$ and $\mathbf{b}(n)$ both and include the data at the sampling time n, so that $\mathbf{w}_{\mathrm{opt}}$ can only be calculated after the sampling data are obtained. In order to be consistent with the nature of the LMS algorithm in Eq. (4.48), that is, use n data of the third sampling point calculates $\mathbf{w}(n+1)$, and we must use Eq. (4.60) to calculate $\mathbf{w}(n+1)$ that the purpose of the RLS algorithm is to make

$$\mathbf{w}(n+1)=\mathbf{A}^{-1}(n)\mathbf{b}(n) \tag{4.61}$$

It is mainly based on the past value $\mathbf{w}(n)$ calculation of the filter coefficient $\mathbf{w}(n+1)$, combined with (4.61), i.e.,

$$\mathbf{w}(n)=\mathbf{A}^{-1}(n)\mathbf{b}(n-1) \tag{4.62}$$

From the definition of Eq. (4.58), we can see that $\mathbf{b}(n)$ could be calculated only through the calculation $b(n-1)$ at the sampling moment n, i.e.,

$$\mathbf{b}(n)=\lambda\mathbf{b}(n-1)+\mathbf{x}(n)d(n) \tag{4.63}$$

Similarly, there is

$$\mathbf{A}(n)=\lambda\mathbf{A}(n-1)+\mathbf{x}(n)\mathbf{x}^{\mathrm{T}}(n) \tag{4.64}$$

And use the inverse of $\mathbf{A}(n)$ calculation $\mathbf{w}(n+1)$. However, even so, calculating the inverse of $\mathbf{A}(n)$ at each sampling instant could be a very arduous task. Thus, using $\mathbf{A}(n-1)$ computations $\mathbf{A}(n)$ becomes very desirable, which could be achieved through the special case of the Woodbury inverse formula—the lemma of the matrix inverse, in the form of

$$\mathbf{A}^{-1}(n) = \lambda^{-1}\mathbf{A}^{-1}(n-1) - \frac{\lambda^{-2}\mathbf{A}^{-1}(n-1)\mathbf{x}(n)\mathbf{x}^{T}(n)\mathbf{A}^{-1}(n-1)}{1+\lambda^{-1}\mathbf{x}^{T}(n)\mathbf{A}^{-1}(n-1)\mathbf{x}(n)} \tag{4.65}$$

Substituting Eqs. (4.65) and (4.63) into Eq. (4.61) yields

$$\begin{aligned}\mathbf{w}(n+1) = {} & \left[\lambda^{-1}\mathbf{A}^{-1}(n-1) - \lambda^{-1}\alpha(n)\mathbf{A}^{-1}(n-1)\mathbf{x}(n)\mathbf{x}^{\mathrm{T}}(n)\mathbf{A}^{-1}(n-1)\right] \\ & \times [\lambda\mathbf{b}(n-1)+\mathbf{x}(n)d(n)]\end{aligned} \tag{4.66}$$

where

$$\alpha(n) = \frac{1}{\lambda + \mathbf{x}^{\mathrm{T}}(n)\mathbf{A}^{-1}(n-1)\mathbf{x}(n)} \tag{4.67}$$

Equation (4.66) expands and combines Eqs. (4.62) and (4.67). After some transformations, the new filter coefficients could be represented by the previous filter coefficients as

$$\mathbf{w}(n+1) = \mathbf{w}(n)+\alpha(n)\mathbf{A}^{-1}(n-1)\mathbf{x}(n)e(n) \tag{4.68}$$

where

$$\mathbf{e}(n) = d(n) - \mathbf{x}^{\mathrm{T}}(n)\mathbf{w}(n) \tag{4.69}$$

Equation (4.68) together with Eqs. (4.65), (4.67), and (4.69) together form the RLS algorithm. The definition of $\mathbf{e}(n)$ in (4.69) is consistent with the form derived in the LMS algorithm. In the LMS algorithm, $\mathbf{e}(n)$ could be calculated from the filter coefficients obtained from the data of the previous sample point, which is the so-called a priori error. For some algorithms, especially those used for adaptive IIR filters, there is also a so-called a posteriori error, i.e., the filter coefficients $\mathbf{w}(n+1)$ are calculated from the new sampling point data, and the error signal is recalculated.

The RLS algorithm needs to perform $O(I^2)$ calculation at each sampling time; that is, it needs I^2 the order calculation, where I is the order of the filter system. In contrast, the LMS algorithm only needs to perform $O(I)$ calculation at each sampling time.

When the signal is stable and the autocorrelation matrix has a wide range of eigenvalues, the RLS algorithm has better convergence performance than the LMS algorithm. The size of the forgetting factor determines the RLS algorithm's emphasis on convergence speed and imbalance. However, it should be noted that when the signal is unstable, the ability of the RLS algorithm and LMS algorithm to track unsteady signals depends on the application. For example, the LMS algorithm is better than the RLS algorithm for tuning signals.

4.5.2 Simulation and Analysis

Using the filtered reference signal and the desired signal, the RLS function in MATLAB could be used to simulate the control effect of the RLS algorithm. Figures 4.9 and 4.10 are the expected signal and the RLS algorithm simulation to obtain the control output signal and the error signal. Comparison between time domain and frequency domain.

For RLS algorithm, the factors that affect the control effect are forgetting factor and filter order. Figure 4.11 shows the influence of the forgetting factor on the error signal when the filter order is $N = 64$. After rotating it by a certain angle (Fig. 4.12), it could be clearly seen that the overestimation of the forgetting factor is not conducive to reducing the error.

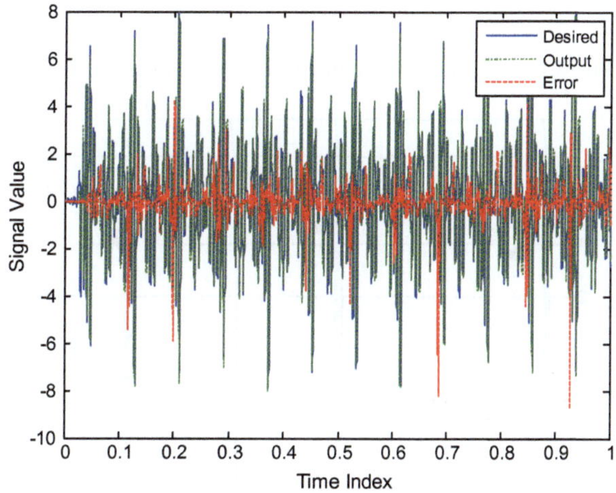

Fig. 4.9 Comparison in time domain among desired signal, control output signal, and error signal obtained by RLS algorithm ($N = 64$, lam $= 0.90$)

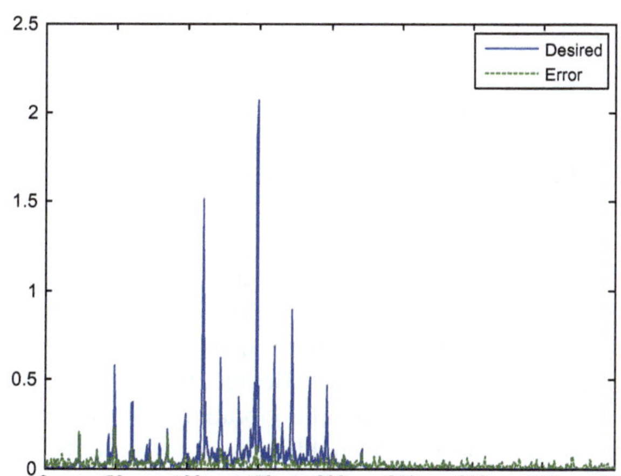

Fig. 4.10 Comparison in frequency domain between desired signal and error signal obtained by RLS algorithm ($N = 64$, lam $= 0.90$)

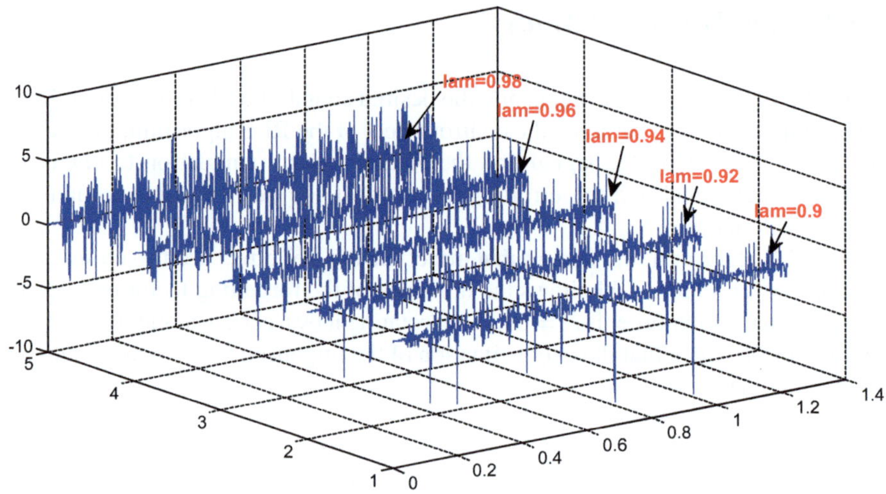

Fig. 4.11 Effect that forgetting factors made on error signals ($N = 64$)

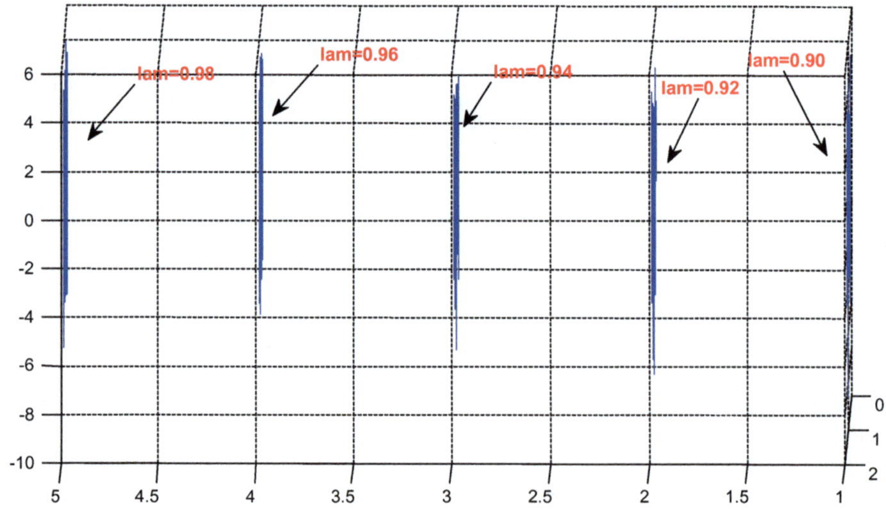

Fig. 4.12 Figure 4.11 rotated by a certain degree

Figure 4.13 shows the influence of the filter order on the error when the forgetting factor $\lambda = 0.95$ is used. Rotate the angle (Fig. 4.14). It could be clearly seen that increasing the filter order blindly does not reduce the error. The filter order is increased to a certain degree. After the degree, the error will become larger.

Figure 4.15 compares the vibration isolation effects of the LMS algorithm and the RLS algorithm. It could be seen from the figure that the control effect of the LMS algorithm below 100 Hz is better than that of the RLS algorithm. The control

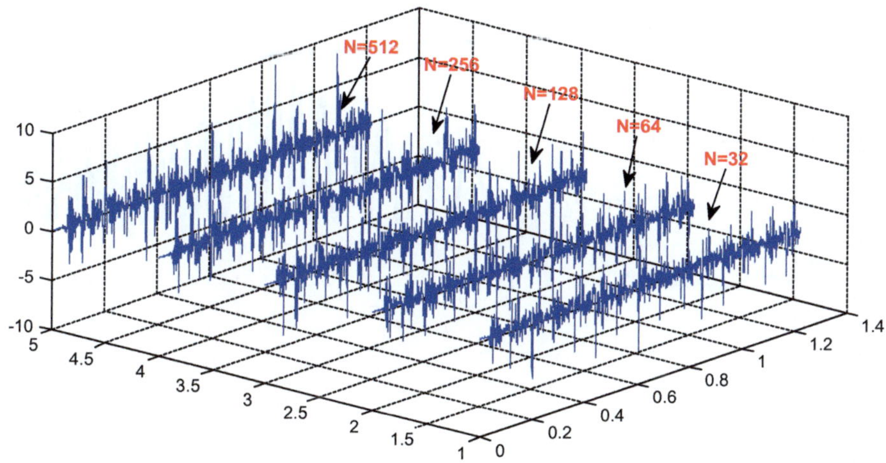

Fig. 4.13 For the forgetting factor lam = 0.96, the effect that filter order made on errors

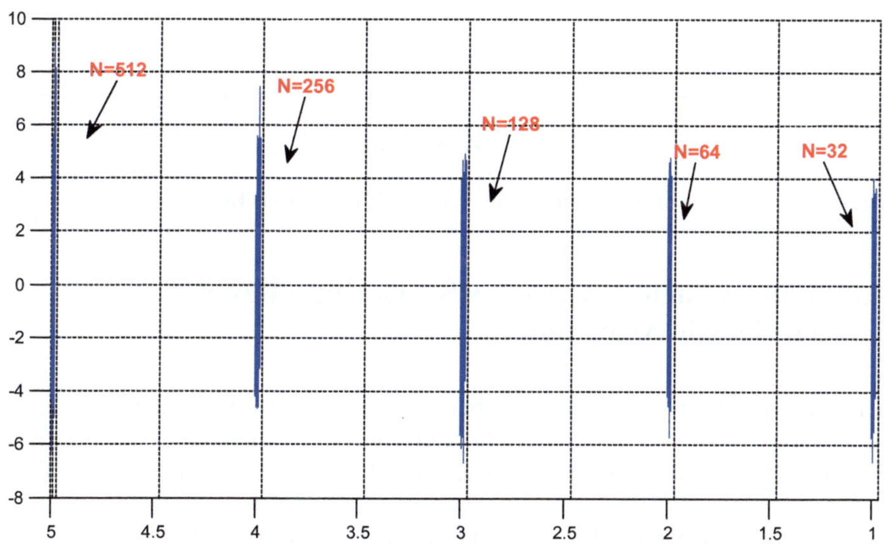

Fig. 4.14 Figure 4.13 rotated by a certain degree

effect of the RLS algorithm above 100 Hz is better than that of the LMS algorithm, and the RLS algorithm needs to be considered at the same time. Due to the long computation time, the LMS algorithm is considered as the control algorithm of the active vibration isolation system.

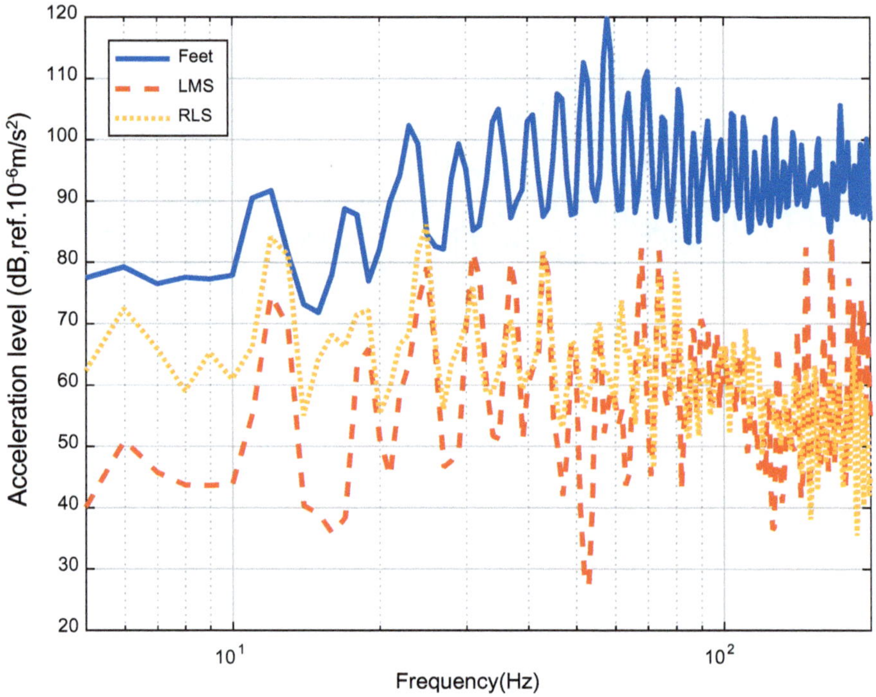

Fig. 4.15 Comparison of control effects between LMS algorithm and RLS algorithm

4.6 Improvement of Leak-LMS Algorithm Based on Genetic Algorithm

The LMS algorithm may cause the filter coefficient to overflow due to the change of operating conditions during operation, thus making the control system unstable. To solve this problem, some experts proposed a Leaky-LMS algorithm with leakage factor.

The Leaky-LMS algorithm performance function is defined as:

$$J(\mathrm{n}) = e^2(n) + \alpha \sum_{i=0}^{N-1} w_i^2(n) \tag{4.70}$$

where α is the leakage factor of the value (0, 1). Due to the existence of the leakage factor, the performance function represented by Eq. (4.70) is different from the standard LMS algorithm. The Leak-LMS algorithm mitigates the problem of coefficient overflow in the standard LMS algorithm because the performance function is not only responsible for $e^2(n)$ but also responsible for the filter coefficients. The formula for updating the adaptive filter coefficients in the Leaky-LMS algorithm is:

$$\mathbf{w}(n+1) = (1 - \mu\alpha)\mathbf{w}(n) + \mu e(n)\mathbf{x}(n) \tag{4.71}$$

At that time $\alpha = 0$, the Leak-LMS algorithm becomes a standard LMS algorithm. A large leakage factor will cause a large steady-state error.

Although the operating conditions must exist, the filter coefficients of the LMS algorithm do not necessarily overflow. Although Leaky-LMS algorithm will eliminate some of the overflow phenomenon, the value of the leakage factor directly affects the performance of the control system. Eliminating the overflow phenomenon, if it is too large, the steady-state error will increase. In this paper, the Leaky-LMS algorithm is improved by referring to the mutation operation in the genetic algorithm, which can effectively solve the above problems.

Prof. J. Holland of the University of Michigan, USA, was inspired by biological evolution to propose GA—genetic algorithm. Genetic algorithm is based on the biological evolution process widely existing in nature, and the concepts of breeding, selection, hybridization, variation, and competition are introduced into the algorithm. Genetic algorithm is essentially a method for efficient global search of problems. It effectively uses existing information to automatically acquire and accumulate knowledge about the search space in the search process and adaptively adjusts the search direction and ultimately achieves an optimal solution.

Genetic algorithms are basically based on a parallel search mechanism, considering many prototype solutions in an arbitrary iteration. All prototype solutions are coded as finite length chromosomes or strings, combined into features or genes. Table 4.1 shows the gene strings in binary.

The genetic algorithm first encodes the solution to the problem; that is, it uses the chromosome to represent possible solutions to the problem and forms the initial population. Common encoding methods include binary encoding, which means that the solution to the original problem is represented as a string consisting of 0s and 1s in binary, as is shown in Table 4.1; decimal encoding, that is, a possible string representation of the problem with a string consisting of 0–9. Real number coding, i.e., the solution of the problem is expressed as a real number, which is different from the first two coding methods. It directly performs the genetic operation in the form of the solution; that is, in the execution, the genetic space is the solution space of the problem, and the chromosome is directly reflects the laws and characteristics of the optimization problem.

Then determine the fitness function, that is, the value that the algorithm strives to maximize. It needs to be determined according to the performance function of the optimization problem. Then, according to the size of the fitness function value, individuals are selected to participate in subsequent genetic operations. The process of mapping the performance function to the fitness function generally requires calibration. However, in the simplest case, the fitness function could be a negative value of the performance function, and then an appropriate performance function and fitness could be calculated for each character string. Calibration of more complex fitness functions could be used to prevent a small number of strings (which may have a considerable value) to dominate the entire selection process, reducing the diversity of subsequent groups.

Table 4.1 Binary representation of gene string

1	0	1	1	1	1	0	0	0	0	1	1	1	0	1	0

In the next iterative process, calibration is also used to increase the diversity of fitness function values in certain groups and encourage convergence. In the most general calibration scenario, the new fitness function is linearly scaled by the old fitness so that the strings at each average fitness function value are only selected once, and those with the greatest fitness values are choose twice.

Finally, according to the biological principles of the survival of the fittest and survival of the fittest, degeneration is performed from generation to generation until an optimal solution to the problem or an approximate optimal solution is obtained.

In each iteration of the genetic algorithm, the strings in this generation, the current population, are used to generate the next generation of strings. The first-generation string is usually selected randomly from all possible combinations of strings. The purpose of the selection operation is to select good individuals from the current population and let them be directly passed on to the next generation as parents or passed on to the next generation in a way that through crossing operation produces new individuals.

After selecting a pair of parent strings, use crossover and mutation to group them together.

Mutation is the process of replacing certain genes in an individual's chromosome code with other genes to form a descendant. For binary encoding, 0 becomes 1 or 1 becomes 0. The mutation operation can overcome the disadvantages of the local search for the cross-operation process to a certain extent and increase the diversity of the population. From the point of view of the ability to produce progeny in the process of genetic evolution, cross-operation is the main method of generating offspring individuals. It directly determines the global search ability of the algorithm. However, although the mutation operation is only an auxiliary method, it is crucial for evolution because it determines the local search ability of the algorithm. Cross-operation and mutation operation cooperate with each other to decide whether the algorithm can converge successfully.

The commonly used mutation operations are:

(a) Basic mutation operation

The basic mutation operation refers to random individuals from the population according to a certain probability and performs mutation operations on the designated locations. For example,

$$X1: 1,\ \ 0,\ \ 0,\ \ 1,\ \ 0,\ \ 1,\ \ \underline{0},\ \ 0$$
$$X1': 1,\ \ 0,\ \ 0,\ \ 1,\ \ 0,\ \ 1,\ \ 1,\ \ 0$$

It is the seventh bit of $X1$ the mutation operation.

(b) Adaptive mutation operation

The adaptive mutation operation and the basic mutation operation process are similar. The only difference is that the probability of mutation varies with the iterative process. Generally, the Hamming distance between two child individuals obtained according to the cross-operation is coded on the corresponding bit of the two character strings. The number of different bits is called the code distance or the Hamming distance. The smaller the Hamming distance, the greater the variation probability; otherwise, the smaller the variation probability.

In Leaky-LMS, μ taking a fixed value in the middle (0, 1), although there is a certain moderating effect on the overflow of the filter coefficient, in practical use, the filter overflow effect will still appear under the change of operating conditions over time. This leads to system instability.

To solve this kind of problem, consider to introduce mutation operation in the genetic algorithm into Leak-LMS, that is, to select partial coefficients $\mathbf{w}(n)$ in a certain probability to perform zeroing operation at regular intervals. In the genetic algorithm, the mutation is to avoid making the genetic algorithm fall into a local optimum and using mutation here can effectively avoid the filter coefficient from overflowing in the case of working condition creep. That is, based on Eq. (4.71), the improved algorithm adds the following steps:

```
{
  For i = 1 : h
    If rand(i) > β
    w(i) = 0
    End
  End
}
```

Among them, the value of rand (i) is between (0, 1); the value of β the same is in among (0, 1), which determines the frequency of the mutation operation. If the value is too small, the algorithm is not easy to converge. If the value of rand (i) is too large, the mutation operation is not easy to be implemented and needs to be performed in the algorithm. The debugging is determined according to the actual situation.

4.7 Hardware Design for Controller

In front of the design of the software in the active control system, the design of the controller and power amplifier in the active control system needs to be completed. In view of the complicated structure of the diesel engine, the vibration of each leg is inconsistent, and it is difficult to use a lumped control method that requires accurate knowledge of the controlled object model. Therefore, a distributed control is selected for the active vibration isolation control of the diesel engine.

According to the selected control mode (distributed control) and control strategy (adaptive feed-forward), it could be seen that the four active and passive hybrid vibration isolator controllers require four controllers. Each controller has the same function and all of them need to have at least two controllers. Acquisition input channels (acquisition of reference signals and desired signals) and an output channel (control signal output), which is an on-chip resource of the controller, should have at least two ADC conversion channels and one DAC output channel, and a micro-processing unit. In addition, as a mature controller, it should also have downloaded debugging and communication functions, as is shown in Fig. 4.16.

In addition, consider the following issues for chip selection:

Selecting the microprocessor needs to focus on the processing speed, memory, and maturity and the help files that could be obtained; therefore, TI's TMS320F28335 dual-core floating-point processor could be selected for this engineering application, with a frequency of up to 150 MHz. The on-chip memory mainly includes two parts: the on-chip RAM storage space is 34 K × 16 bit, and the on-chip FLASH storage space is 64 K × 16 bit. After the algorithm debugging is completed, the DSP program is solidified in the on-chip FLASH memory, and it is automatically loaded into the internal RAM memory for execution after being powered on.

Selecting the ADC and DAC requires attention to conversion speed, precision, and number of channels. The overall result of the selection is:

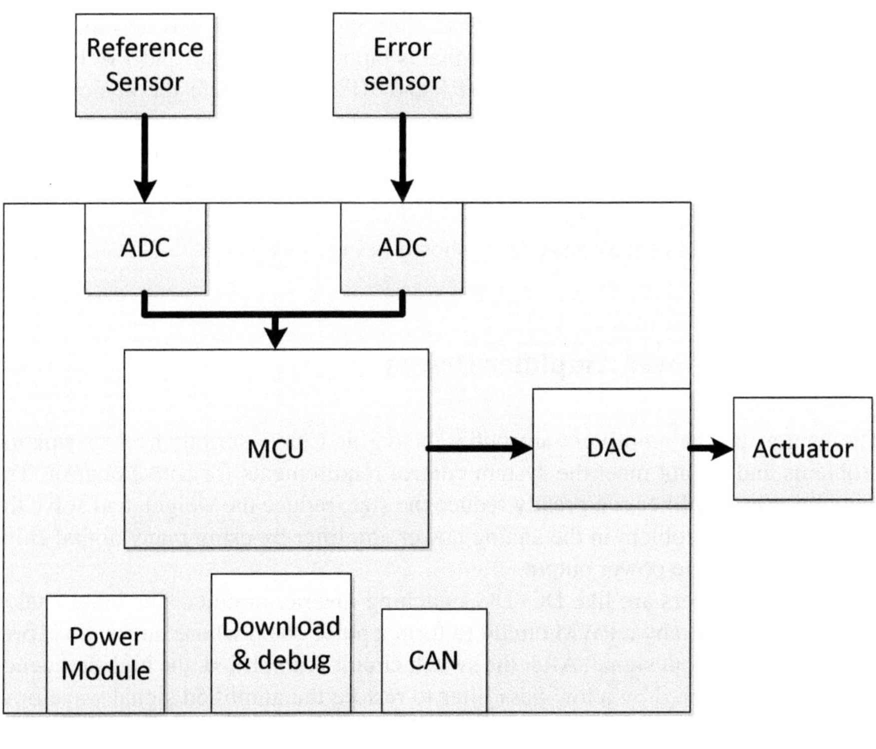

Fig. 4.16 Design schematic of distributed adaptive feed-forward controller

Fig. 4.17 Prototype of distributed adaptive feed-forward controller

ADC: TI's 16-bit high-accuracy ADC chip, the ADS8555, has an input signal range of ±10 V. The external analog signal is input to channels 0 and 1 of the ADC device through the BNC interfaces XP1 and XP2 to complete the analog signal conversion.

DAC: TI's 16-bit high precision DAC chip DAC8718 with an output signal range of ±10 V. The analog signal converted by Channel 0 of the DAC device is output through the BNC interface XP3.

The final design of the controller is shown in Fig. 4.17.

4.8 Digital Power Amplifier Design

The analog power amplifiers are bulky, heavy and have serious heat dissipation problems and do not meet the system control requirements for active control. The digital power amplifier can greatly reduce the size, reduce the weight, and solve the heat dissipation problem in the analog power amplifier by using many digital chips to provide the same power output.

Digital amplifiers are like DC–DC switching inverter circuits. The input analog signal is modulated by a PWM circuit to form a pulse chain whose duty ratio is proportional to the input signal. After the switch circuit is amplified, the high-frequency component is filtered by a low-pass filter to restore the amplified signal waveform. Output.

A typical digital amplifier consists of a display panel, an auxiliary power supply, and two power-rate modules. Among them, 220 V AC power supply is used to supply two power modules and auxiliary power supply. The auxiliary power supply

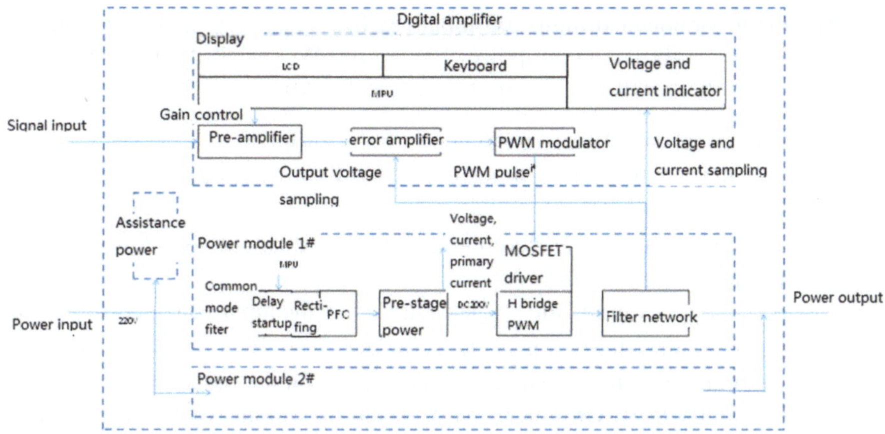

Fig. 4.18 Schematic of digital power amplifier

provides low voltage power supply ±15 and ±5 V for the display panel and two power modules. After the 220 V AC power is sent to the power module, it is filtered by the common mode filter to improve the EMC performance. Then the surge current is controlled by the slow start circuit, followed by rectification, power factor correction (PFC), and 200 V DC output from the front stage power supply. Power supply to the rear amplifier. After the analog signal is input, it is amplified by the preamplifier of the control board, and the signal gain is adjusted and sent to the error amplifier. After the input reference signal is subtracted from the output voltage sampling signal of the power module, an error signal is output and sent to the PWM modulator to generate PWM. The driving pulse is driven by the MOSFET driver to drive the H-bridge and the modulated power signal is output, and the output power is low through the filter network, as is shown in Fig. 4.18.

4.8.1 Hardware Design of Digital Amplifier

According to the previous test results of the DC resistance and static thrust of the actuator, it could be known that when the current is input at 5 A, a power of 1500 N could be output. At this time, the power demand is $I^2 R = 5 \times 5 \times (4.334 + 4.230 + 4.139 + 4.275) = 424.45$ W. Consider leaving a certain margin to set the power output of the power amplifier as 500 W.

(1) PFC

The power factor correction circuit adopts the single-cycle control mode. The control chip uses Infineon's ICE3PCS02G. This method requires neither a multiplier nor an input voltage to detect, which greatly simplifies the circuit design.

The main parameter design: AC input (RMS): $V_{\text{in}} = 176–264$ V; output DC voltage: $V_{\text{o}} = 400$ V; output power: $P_{\text{o}} = 1200$ W; switching frequency: $f_{\text{s}} = 100$ kHz; efficiency: $\eta = 92\%$; power factor: PF = 0.99.

(1) Numerical calculation of PFC boost inductor L
The maximum current input by the grid is: $I_{\text{in(RMS)MAX}} = \frac{P_{\text{inMAX}}}{V_{\text{inMIN}}\text{PF}} = \frac{P_{\text{o}}}{V_{\text{inMIN}}PF\eta} = 7.49\,\text{A}$.
The maximum peak current of the grid input is: $I_{\text{in(PK)MAX}} = \sqrt{2}I_{\text{in(RMS)MAX}} = 10.59\,\text{A}$.
Since the maximum duty cycle occurs when the input voltage is minimum, the maximum duty cycle is: $D = \frac{V_{\text{o}} - V_{\text{in(PK)MIN}}}{V_{\text{o}}} = \frac{V_{\text{o}} - \sqrt{2}V_{\text{in(PK)MIN}}}{V_{\text{o}}} = 0.38$.
Assuming that the maximum ripple current on the inductor is 20% of the maximum input current to the grid, the maximum ripple current on the inductor is: $\Delta I = 0.2I_{\text{in(PK)MAX}} = 2.12\,\text{A}$.
The PFC boost inductor L is: $L = \frac{V_{\text{in(PK)MIN}}D}{f_{\text{s}}\Delta I} = 446\,\mu\text{H}$.
Due to the influence of the exciting current of the inductor, the magnetic permeability has decreased, so taking the inductance value slightly larger than the calculated value is 560 μH.

(2) Selection of power semiconductors
The maximum current flowing through the switch is $I_{\text{in(PK)MAX}} + \frac{\Delta I}{2} = 11.65\,\text{A}$, choose Infineon's SPW20N60C3, the MOSFET can flow 13.1 A current at 100 °C.
The step-up diodes were selected from the American company Thomas's silicon carbide diode CI20S65. When the MOS tube is below 135 °C, the allowable passing current is 28 A to meet the requirements. Withstanding voltage of 650 V, with zero recovery characteristics, that is, the reverse recovery current time is less than 20 ns, almost no reverse recovery current, which can greatly increase the operating frequency and work efficiency. It is not necessary to absorb the circuit in the PFC, which can greatly simplify the circuit design. Improve circuit reliability and reduce EMI.

(2) Pre-stage power supply
The pre-stage power supply adopts UC3875 as the control chip, the input voltage is DC 400 V, the highest output voltage is DC 200 V (DC control), and the output power is 1100 W. Full-bridge phase-shift controller UC3875.
Power transformer design:
Switching frequency 100 kHz, $T_{\text{s}} = 10\,\mu\text{s}$ dead zone $t_{\text{d}} = 0.1\,\mu\text{s}$, maximum pulse width $T_{\text{on}}\max = \frac{T_{\text{s}}}{2} - t_{\text{d}} = 4.9\,\mu\text{s}$.
The PQ5050 ferrite core with PC40 material is used.
The primary number of transformers is: $N = \frac{100Et_{\text{on}}}{2S_{\text{c}}B_{\text{m}}} = 31$, where E is the input voltage, t_{on} is the pulse width, and B_{m} is the maximum magnetic induction intensity during operation is 1000 Gs, $S_{\text{c}} = 3.14\,\text{cm}^2$ is the effective cross-sectional area of the magnetic flux.

The primary inductance $L = \frac{\mu_0 \mu_r N^2 S_c}{l_c} = 8.39\,\text{mH}$, $\mu_0 = 4\pi \times 10^{-9} H/m$ is the vacuum permeability, μ_r is the relative permeability, and the value is 2500, $l_c = 11.3\,\text{cm}$ is the length of the magnetic circuit.
Excitation current is: $I_\mu = \frac{E t_{on}}{L} = 0.234\,\text{A}$.
The secondary number of turns is: $N_n = \frac{V_o N}{E} = 16$, and $V_O = 200\,\text{V}$ is output voltage.

(3) H-bridge PWM
Using four MOSFETs to form the H-bridge, which was used to chop 200 V DC voltage to output the modulation waveform, the operating frequency of the H-bridge is 200 kHz, and Infineon's SPW20N60C3 is selected.
Due to the high switching frequency of the H-bridge, the IR2181 chip from International Rectifier was used for MOSFET driving. The IR2181 drives an output current of up to 1.9 A to drive the SPW20N60C3.

(4) Filtering network
The PWM signal output by the H-bridge needs to be low-pass filtered before it could be restored to the desired waveform. The FilterSolutions software is used to assist the design. The parameters are Fourth-order LC Butterworth low-pass filter, cut-off frequency is 10 kHz, and the output is the matching impedance is 5 Ω. The output parameters are shown in Fig. 4.19.

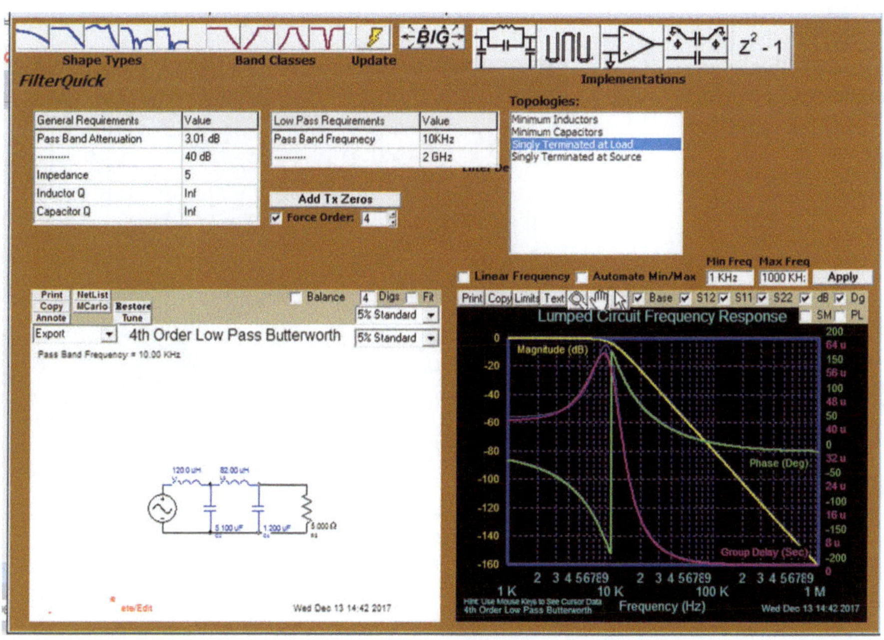

Fig. 4.19 Screenshot of software design for digital power amplifier

4.8.2 Performance Test

According to the designed hardware processing and assembly power amplifier, the power amplifier prototype is shown in Fig. 4.20. After the assembly, the performance parameters are tested. The main test contents include voltage, current, frequency response, nonlinear distortion, signal-to-noise ratio, and power. Factor test.

Amplifier voltage and current measurement methods are: signal source output sine signal to digital amplifier, amplifier output access ammeter and voltmeter measurement of current and voltage, the load using resistive load, as is shown in Fig. 4.21.

Among them, when measuring the maximum output voltage, use a 10 Ω resistor load, set the digital power amplifier gain to 25, monitor the output voltage with an oscilloscope (note that the oscilloscope must be left unconnected and must not be grounded), and adjust the output of the signal source so that the voltage on the load is The maximum is achieved without distortion, and the voltmeter display voltage is the maximum output voltage. In theory, it should be greater than 100 V, and the actual measurement result is 105 V, to meet the design requirements.

When measuring the maximum output current, use a 2.5 Ω resistor load, set the digital power amplifier gain to 25, increase the protection current, monitor the output voltage with an oscilloscope (note that the oscilloscope must be left floating, must not be grounded), and increase the output of the signal source to make the load.

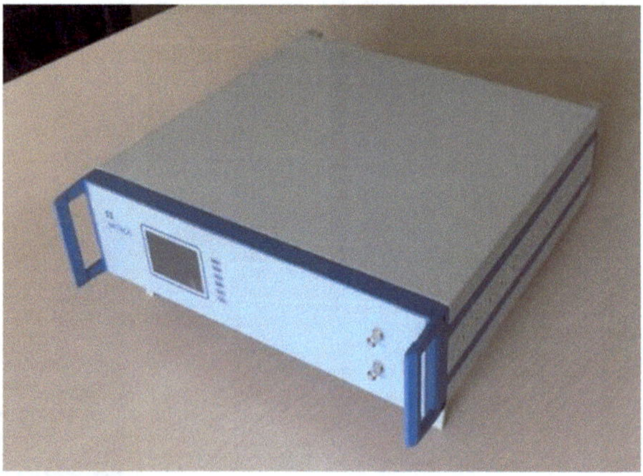

Fig. 4.20 Prototype of digital power amplifier

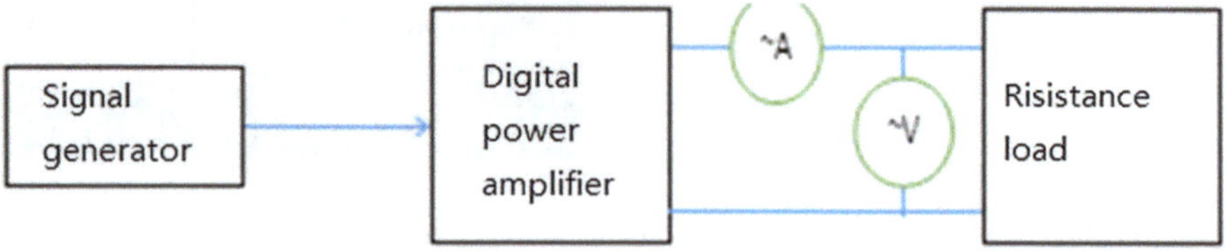

Fig. 4.21 Schematic of voltage and current test

The current increases; in theory, the ammeter should be able to reach 20 A and the measured value >20 A.

The test method of the frequency response of the power amplifier is as follows: The signal source outputs a sine signal to the digital amplifier, the output of the power amplifier is connected to the current meter and the voltage and the electric current are measured, and the load is a resistive load, as is shown in Fig. 4.22.

The specific operation is as follows: Using 5 Ω resistor load, the signal source is set to 1 V (RMS) output, and the digital amplifier gain is set to 20. Change the frequency of the signal source and measure the frequency response, within 500 Hz, the voltage change should not exceed 5%. The specific test results are listed in Table 4.2. It could be seen that the design requirements could be met.

The test method for nonlinear distortion is as follows: The signal source outputs a sine signal to the digital amplifier, the output of the amplifier is connected to a resistive load, and the voltage on the load is connected to a distortion tester (DF4120), as is shown in Fig. 4.23.

The specific operation is as follows: Using 5 Ω resistor load, the signal source is set to 1 V (RMS) output, the frequency is 150 Hz, the gain of the digital power amplifier is set to 20, the measurement distortion should be less than 1%, the measured 0.6%, to meet the design requirements.

The signal-to-noise ratio test method is as follows: The signal source outputs a sinusoidal signal to a digital amplifier, the load is a resistive load, and the output voltage is measured with the AC voltage profile of multimeter, as is shown in Fig. 4.24.

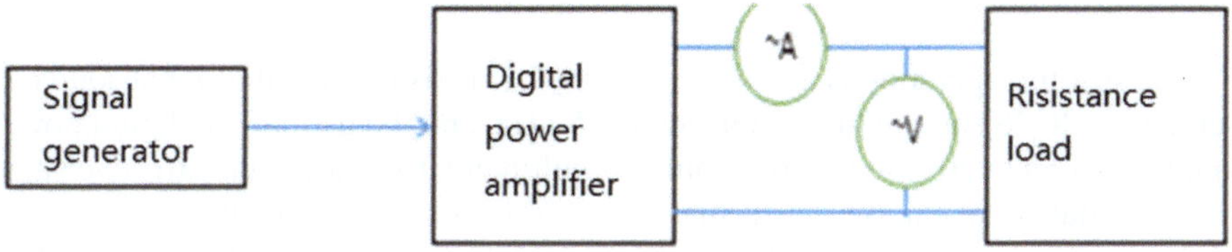

Fig. 4.22 Schematic of frequency response test

Table 4.2 Results of frequency response test

Frequency (Hz)	50	100	200	300	400	500
Output voltage (V)	15.3	15.3	15.3	15.2	15.1	15.2
Maximum (V)	15.3					
Minimum (V)	15.1					
Relative error	1.3%					

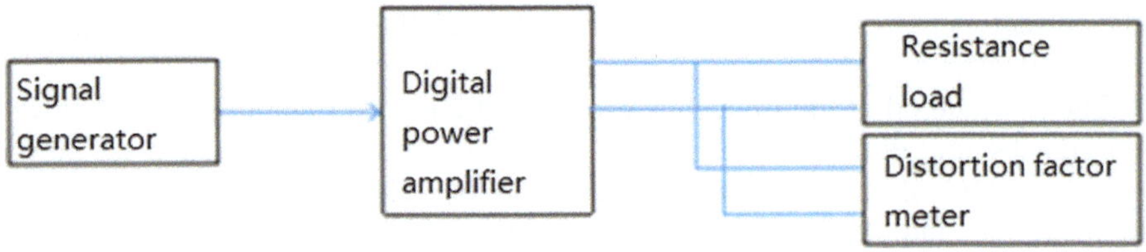

Fig. 4.23 Schematic of nonlinear distortion test

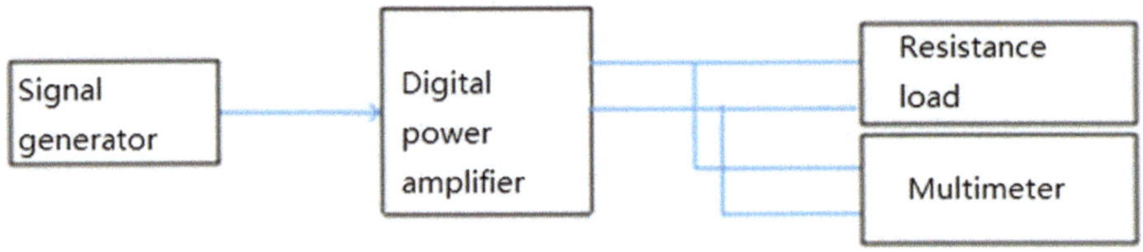

Fig. 4.24 Schematic of signal-to-noise ratio test

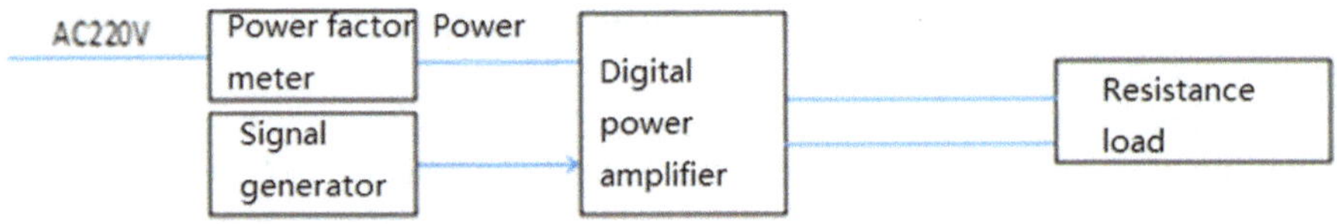

Fig. 4.25 Schematic of power factor test

The specific operation is as follows: Using 5 Ω resistor load, the signal source is set to 0 V (RMS) output, the frequency is 150 Hz, and the gain of the digital power amplifier is set to 20. Use the AC profile of multimeter to measure the effective value of the output voltage noise. The actual value is 0.02 V. The maximum output voltage of the amplifier is 100 V, and the output signal-to-noise ratio is 20log (100:0.02) = 74 dB. It could be seen that the designed power amplifier has a high signal-to-noise ratio.

The test method for the power factor is as follows: The signal source outputs a sinusoidal signal to the digital amplifier, and the load is a resistive load. A power factor meter is inserted at the power end of the power amplifier, as is shown in Fig. 4.25.

The specific operation is as follows: Using 5 Ω resistor load, the signal source is set to 3 V (RMS) output, the frequency is 150 Hz, the gain of the digital power amplifier is set to 20, the reading of the power factor meter theoretically should be greater than 0.95, and the actual measurement is 0.96, which meets the design requirements.

4.9 Conclusion

In this chapter, based on the characteristics of active vibration isolation of diesel engines, an adaptive feed-forward control algorithm is studied. An adaptive feed-forward control method combining the characteristics of active and passive hybrid vibration isolation is proposed. The control effect of LMS algorithm and RLS algorithm is simulated and analyzed using measured data. The hardware systems such as power amplifiers and controllers are designed to solve the problems of large, heavy, and heat-dissipating analog power amplifiers. Research shows:

(1) Based on the MATLAB platform and measured diesel vibration data, the simulation control effect of LMS and RLS algorithm shows that the control effect of LMS below 100 Hz is better than that of RLS, and the control effect of RLS above 100 Hz is better than that of LMS; in addition, it takes longer to consider RLS. In addition, RLS takes much more time to operate, so LMS algorithm is selected as the active vibration isolation system control algorithm. At the same time, the Leaky-LMS algorithm is improved based on the genetic algorithm, and the filter coefficient overflow phenomenon that may occur due to the change of operating conditions in the LMS algorithm is solved. Therefore, suitable control algorithms could be determined according to the application scenarios, and the optimal parameters for the controller in the active–passive hybrid vibration isolation of the diesel engine could be determined.

(2) Based on the selected control method and control strategy, combined with the characteristics of diesel active vibration isolation system, a distributed adaptive feed-forward controller was designed and developed. A complete diesel active vibration isolation system requires four controllers. Each controller contains two ADC conversion channels, which are used to acquire reference signals and error signals, respectively. A DAC conversion channel is used to output control signals to drive the actuator.

(3) A digital power amplifier with a rated output power of 500 W is designed to meet the power requirements and operating characteristics of the designed active–passive hybrid isolator, which solves the problems of large, heavy, and heat-dissipating analog power amplifiers. Through tests on voltage, current, frequency response, nonlinear distortion, signal-to-noise ratio, and power factor, the output voltage of the amplifier is 105 V, output current is > 20A, voltage relative deviation is 1.3%, and nonlinear distortion is 0.6%, signal-to-noise ratio is 74 dB, and power factor is 0.96, which meets the application requirements of diesel engine active and passive hybrid vibration isolation.

Reference

1. van der Sande TPJ (2011) Control of an automotive electromagnetic suspension system. Eindhoven University of Technology

Chapter 5
Comprehensive Experimental Verification for AVI

Abstract This chapter studies the application technology of active and passive hybrid vibration isolation. The active and passive hybrid vibration isolator designed in Chap. 2 and the active control system designed in Chap. 3 are applied to the control of diesel low frequency vibration transmission. Firstly, the vibration characteristics of diesel engine under typical installation environment and the influence of diesel engine noise on base vibration are analyzed. Based on passive vibration isolation performance test, active control is started to test the effect of active and passive combined control of low frequency vibration transmission under typical operating conditions.

5.1 Introduction

The first three chapters are based on the Halbach magnetic array and develop the active and passive hybrid isolators for marine diesel engines. Based on the characteristics of active and passive vibration isolation of diesel engines, the software and hardware of the active and passive vibration isolation systems are designed and implemented based on the feed-forward control method. The simulation analysis and improvement of the adaptive feed-forward control algorithm, the high-performance distributed feed-forward controller, and the development of a 500 W digital power amplifier were completed.

Although using the measured data to simulate the control performance of LMS and RLS algorithms and solve the overflow problem of LMS filter coefficients, it is proposed that the Leaky-LMS algorithm is improved by the mutation operation in the genetic algorithm. However, irremovable time lags in the control system, changes in operating conditions, actuator response delays, and electromagnetic nonlinearity all affect the final vibration isolation effect; therefore, it is necessary in practical applications. The effectiveness of the vibration isolation of the active and passive vibration isolation system is verified. At the same time, to compare and analyze the control effect of the active and passive hybrid vibration isolation, the vibration

F. Wang et al., *Comprehensive Investigation on Active-Passive Hybrid Isolation and Tunable Dynamic Vibration Absorption*, Springer Tracts in Mechanical Engineering, https://doi.org/10.1007/978-981-13-3056-8_5

characteristics of the diesel engine under typical installation environment must be measured and analyzed.

This chapter will measure and analyze the diesel engine, foot, base, and deck vibrations in a typical installation environment in a large-scale model; measure and analyze diesel engine, machine foot, foundation, and deck vibration caused by diesel engine noise. The vibration isolation effectiveness of active–passive vibration isolator with active controller on and off are verified by experiments.

5.2 Vibration Isolation Performance of AVI Without Control

Replace the WH400 isolator with the active and passive hybrid isolator designed in Chap. 2, as is shown in Fig. 5.1, and measure the vibration at the base position at different rotational speeds.

Use the sensor installation method as is shown in Fig. 5.2 to arrange the sensors to measure the vibration of the diesel engine, the machine foot, the base, and the deck. The relationship between the sensor serial number and the placement position is shown in Table 5.1.

At this time, the comparison of vibration isolation effects of different vibration isolation methods at different speeds is shown in Figs. 5.3 and 5.4, respectively. Among them, AVI-Stiffness indicates that only spring isolation is used, and AVI-Stiffness

Fig. 5.1 Way to install active and passive hybrid vibration isolator

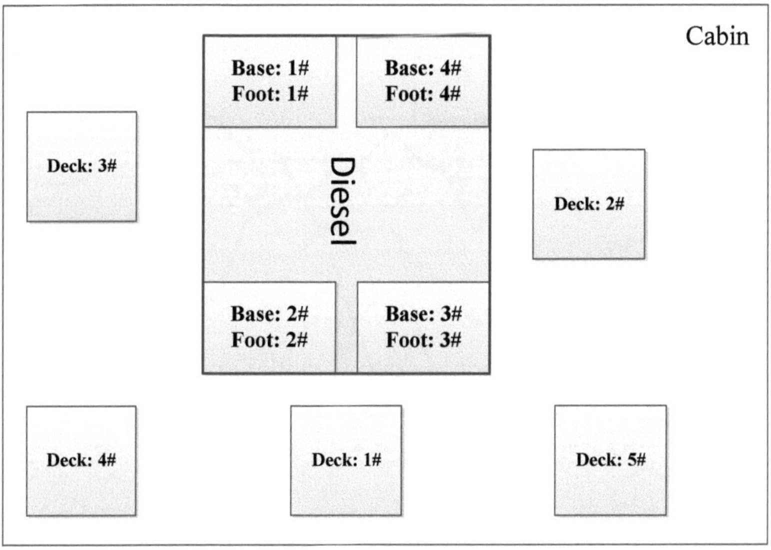

Fig. 5.2 Arrangement of sensors

Table 5.1 Relationship between numbering and locations of sensors

No	Location	No	Location
1	Base No. 1	9	Floor No. 2
2	Base No. 2	10	Foot No. 1
3	Base No. 3	11	Foot No. 2
4	Base No. 4	12	Foot No. 3
5	Vertical of diesel	13	Foot No. 4
6	Longitudinal of diesel	14	Floor No. 3
7	horizontal of diesel	15	Floor No. 4
8	Floor No. 1	16	Floor No. 5

Damping indicates that the spring + eddy current damping vibration isolation. It could be found:

1. Compared with active and passive hybrid isolators with no active control, rubber isolators have excellent vibration isolation in the frequency band from 15 to 600 Hz, with an average of 15 dB.
2. When active control is not enabled, the active and passive hybrid vibration isolator is within 15 Hz. In the frequency band above 600 Hz, it has a better vibration isolation effect than the rubber isolator with an average of 5 dB.
3. For this active–passive hybrid vibration isolator, damping has little effect on vibration attenuation.

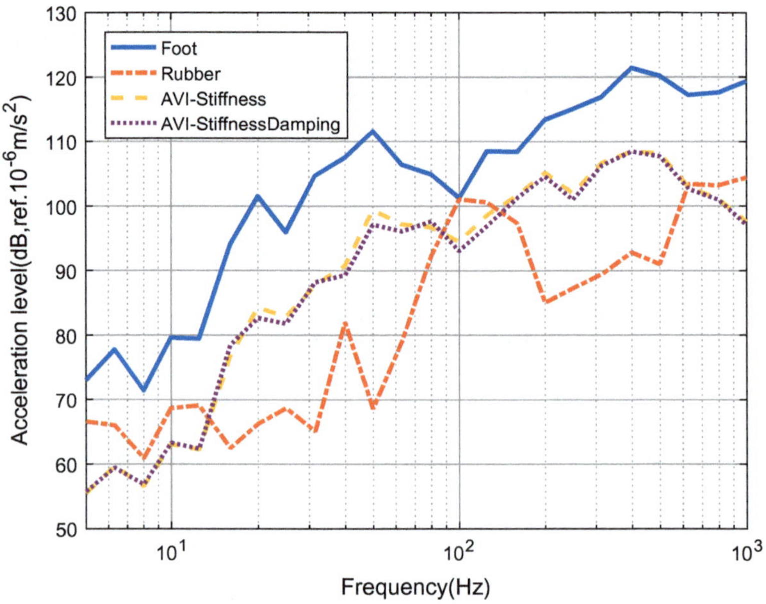

Fig. 5.3 Comparison of vibration isolation effects of different vibration isolation methods of base for the idle rotating speed

The effect of different vibration isolation methods on deck vibration at 1500 RPM is shown in Fig. 5.5. It could be found that the rubber isolator has better vibration isolation performance than the active–passive hybrid isolator without active control.

The effects of different vibration isolation methods on the vertical, lateral, and longitudinal vibration of a diesel engine are shown in Figs. 5.6, 5.7, and 5.8, respectively. It could be found that:

1. For vertical vibration of a diesel engine, installing a rubber isolator reduces vibration by an average of 15 dB in a frequency band below 80 Hz compared to the installation of a passive active isolator.
2. The installation of rubber isolators and the installation of active and passive hybrid isolators have little impact on the lateral vibration of diesel engines; however, for diesel longitudinal vibrations, the installation of active and passive hybrid isolators is more than 100 Hz above the installed rubber isolators. Inside, the vibration decreases more as the frequency increases.
3. The vertical vibration of the diesel engine is like that of the longitudinal vibration, but both are smaller than the magnitude of the lateral vibration. This is because the cylinders of the diesel engine are arranged in a V-shape, and a periodic lateral force is generated during the operation of the diesel engine.

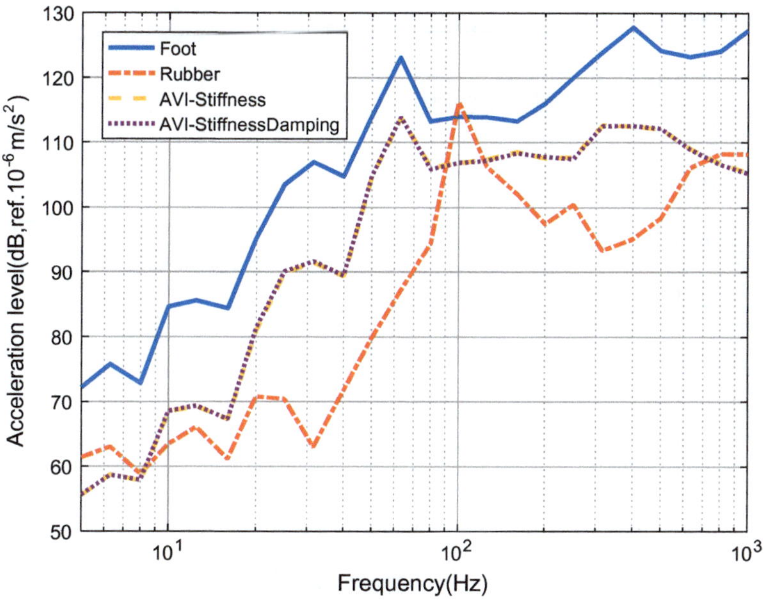

Fig. 5.4 Comparison of vibration isolation effects of different vibration isolation methods of base for the rotating speed of 1500 RPM

5.3 Acoustic-Induced Vibration of Diesel Engine

5.3.1 Causes of Measure Acoustic-Induced Vibration

Considering that diesel engine noise causes base vibration and thus reduces the vibration isolation effect of the mechanical isolation system, to correctly quantify the vibration isolation effect of actively controlled active and passive hybrid vibration isolation, it is necessary to quantify the vibration caused by diesel engine noise.

5.3.2 Acquiring Diesel Engine Noise

The instruments used to collect diesel noise are listed in Table 5.2 (Fig. 5.9 is sound level meter).

The collected sound pressure is listed in Table 5.3.

The corresponding relationship between sound pressure level and frequency under different working conditions is shown in Figs. 5.10, 5.11, 5.12, and 5.13. As could be seen from the figures:

(1) The noise problem of the line spectra corresponding to the engine speed and its half-order harmonics is most prominent. For example, for 1500 RPM (25 Hz/s),

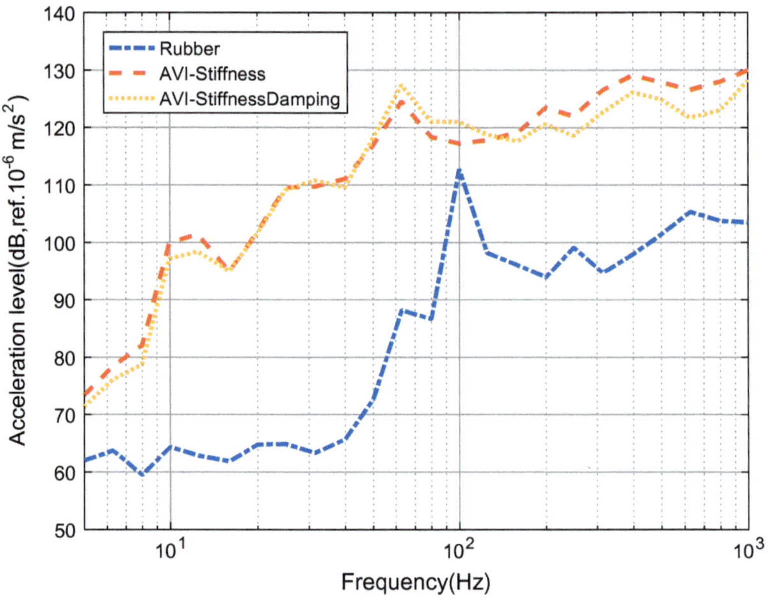

Fig. 5.5 Comparison of vibration isolation effects of different vibration isolation methods of deck for the rotating speed of 1500 RPM

Table 5.2 Instruments used to collect noise from diesel engine

No	Name	Amount	Purpose
1	Data acquisition instrument	1	Collect and store diesel noise data
2	Microphone	1	Measure diesel noise
3	Sound level meter	1	Measurement noise decibels

Table 5.3 Sound pressure corresponding to different working conditions

RPM	Decibel (dB)
Idle speed	85.5
1000	90.2
1200	92.6
1500	95.4

the noise frequencies are 12.5, 25, 37.5, 50 … Hz, respectively. The situation in 4.22 is consistent.

(2) The noise energy is mainly concentrated on low frequencies below 100 Hz.

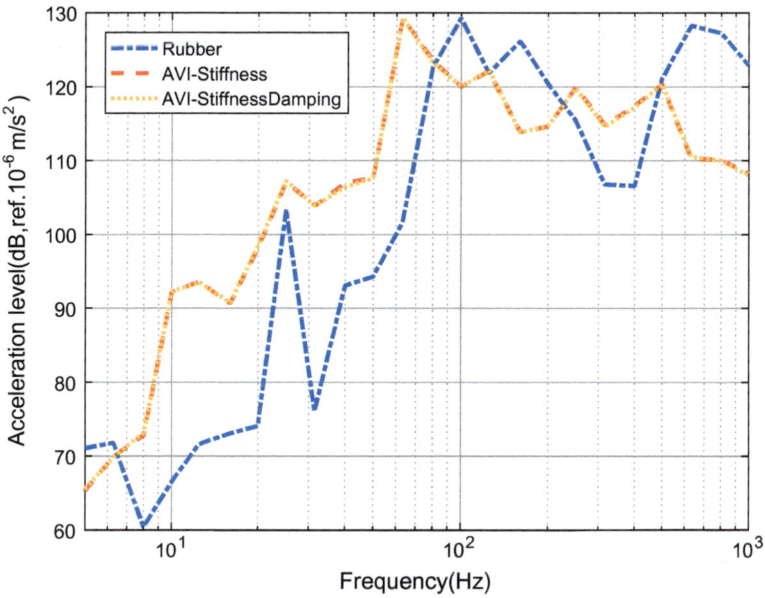

Fig. 5.6 Comparison of vertical vibration levels of diesel engine of different vibration isolation methods for the rotating speed of 1500 RPM

Table 5.4 Instruments used to measure acoustic excitation

No	Name	Amount	Purpose
1	Data acquisition instrument with output function	1	Output diesel noise and collect vibration and noise data
2	Power amplifier	1	Amplify power
3	Loudspeaker	1	Reproduce diesel noise
4	Sound level meter	1	Used to adjust the power magnification to get the same decibels

5.3.3 Acoustic-Induced Vibration Analysis

The collected vibration data are converted from a .dat file into a .wav file, and the noise of the diesel engine is reproduced through the playback device, spherical speaker (Fig. 5.14), and power amplifier (Fig. 5.15), Vibration data.

The instruments used to reproduce the diesel engine noise are listed in Table 5.4.

The comparison between the noise spectrum generated by the diesel engine and the noise frequency generated by the loudspeaker is shown in Figs. 5.16, 5.17, 5.18, and 5.19. It could be found that although they have the same energy, similar waveforms

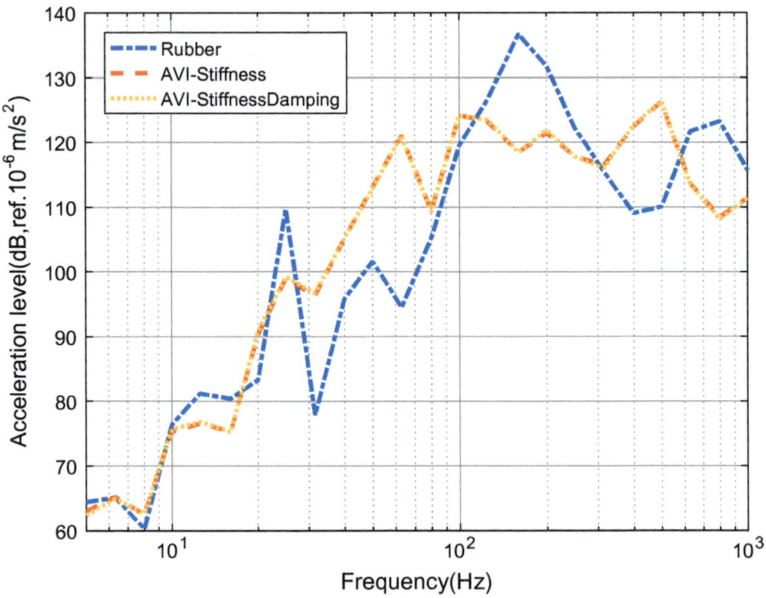

Fig. 5.7 Comparison of lateral vibration levels of diesel engine of different vibration isolation methods for the rotating speed of 1500 RPM

are generated; however, the energy of the sound produced by the loudspeakers is mainly concentrated above 50 Hz, which means that the combination of the power amplifier and the loudspeaker cannot completely reproduce the noise of the diesel engine below 50 Hz.

Through Figs. 5.20, 5.21, 5.22, 5.23, 5.24 and 5.25, we can find:

1. The diesel engine noise has a significant influence on the frequency band below 30 Hz for the base and deck vibrations. It is equal to or close to the vibration of the active and passive hybrid isolators when the diesel engine is opened and is greater than the vibration when the rubber isolator is installed; therefore, if base and deck vibration are the control targets, the control frequency should be higher than 30 Hz.
2. The noise of diesel engine has little effect on the vibration of the diesel engine itself and the foot. If the vibration of the diesel engine is the control target, it could be performed in the frequency range higher than 10 Hz; the effect on the vertical vibration of the diesel engine is less than that of the diesel engine and the effect of longitudinal vibration.
3. In addition, considering the noise of the diesel engine, in the frequency range below 40 Hz, the amplitude of the noise spectrum is lower than the amplitude of the noise spectrum when the diesel is turned on. That is, if the vibration caused by the reproduced noise is taken into account, the vibration caused by the noise within 40 Hz will be close to or higher than the vibration caused by the diesel

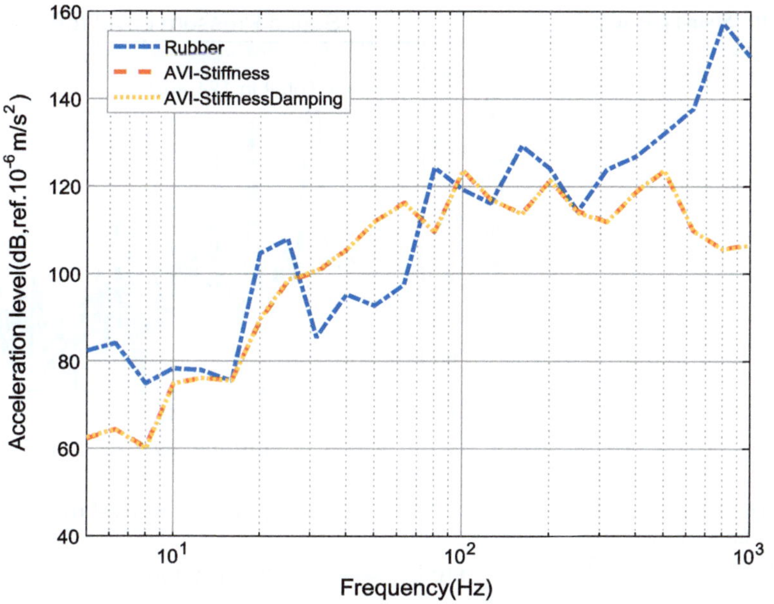

Fig. 5.8 Comparison of longitudinal vibration levels of diesel engine of different isolation methods for the rotating speed of 1500 RPM

Fig. 5.9 Sound level meter

engine being turned on. The vibration caused by the recurrence of diesel noise in the frequency band above 40 Hz will be reduced due to the correction, further lower than the vibration caused by the diesel engine being turned on. Therefore, the control frequency should be higher than 40 Hz, and the true control effect could be observed.

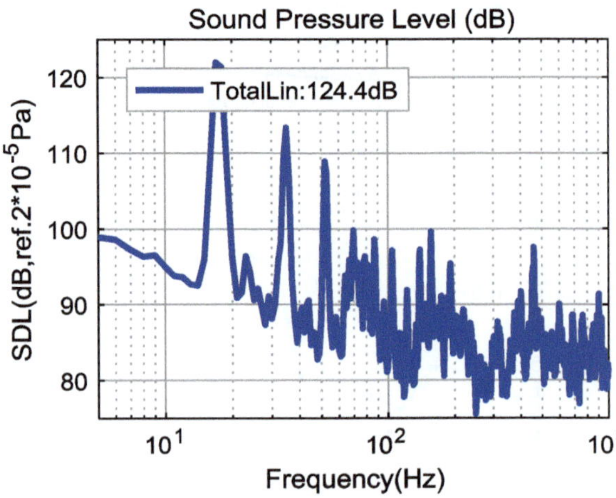

Fig. 5.10 Sound pressure level of diesel engine at the idle rotating speed

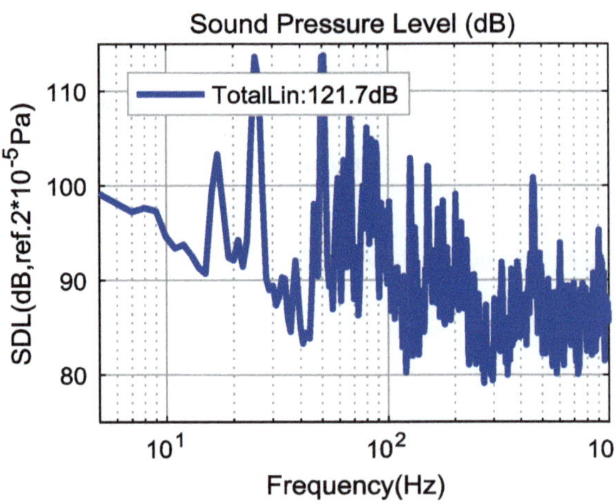

Fig. 5.11 Sound pressure level of diesel engine at the rotating speed of 1000 RPM

5.4 Research on Performance of Active and Passive Hybrid Vibration Isolation for Diesel Engines

5.4.1 Test Method

The test plan for active and passive hybrid vibration isolation performance is shown in Fig. 5.26. Using distributed adaptive feed-forward control, the vertical vibration and deck vibration of the diesel engine are used as the reference source and the error source, respectively. The sensor installation method is shown in Fig. 5.2 and remains unchanged.

The vertical vibration and deck vibration of the diesel engine are measured by an acceleration sensor and then entered a DSP controller through a signal conditioning

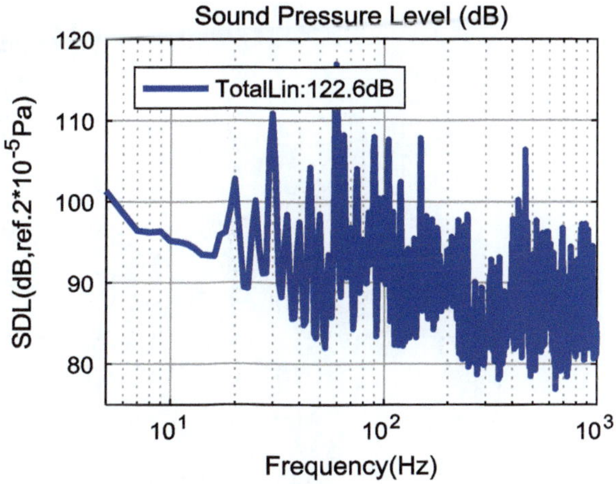

Fig. 5.12 Sound pressure level of diesel engine at the rotating speed of 1200 RPM

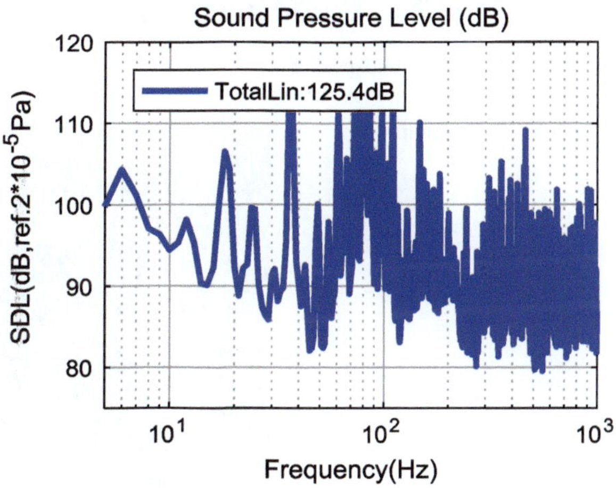

Fig. 5.13 Sound pressure level of diesel engine at the rotating speed of 1500 RPM

circuit. The controller outputs a control signal by running a certain control algorithm and drives the electromagnetic actuator in the isolator after being amplified by the power amplifier. Actuation controls low frequency vibrations from the diesel engine to the deck.

Among them, the signal conditioning circuit is used to amplify the measurement signal and filter out noise while powering the sensor.

The test site is shown in Fig. 5.27.

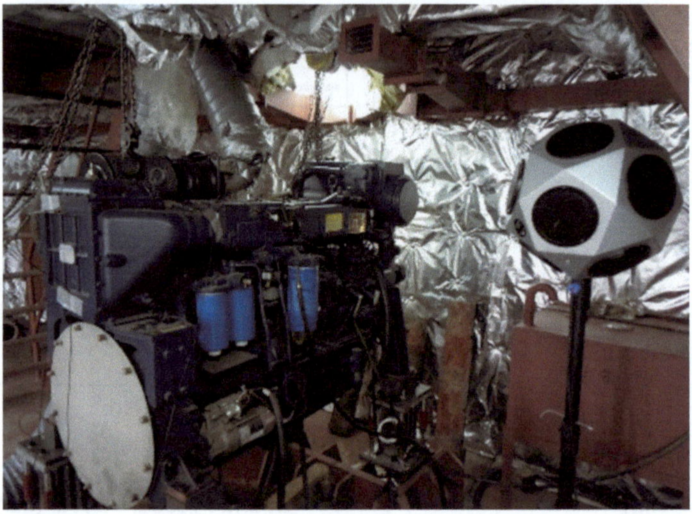

Fig. 5.14 Spherical loudspeaker

Fig. 5.15 Power amplifier

5.4.2 Test Results and Analysis

According to the previous analysis results, the active control frequency band is set at 40–100 Hz, the data collected by the acquisition card are filtered by a 40–100 Hz band-pass filter, and a low-frequency line spectrum adaptive control strategy is applied. The diesel engine and the diesel engine are the vibration of the foot, isolator base, and deck is shown in Figs. 5.28, 5.29, 5.30, and 5.31, respectively. It could be seen that the vibration of the diesel engine at this time is increased at some frequencies. In the frequency band below 20 Hz, the vibration amplification is obvious, with an

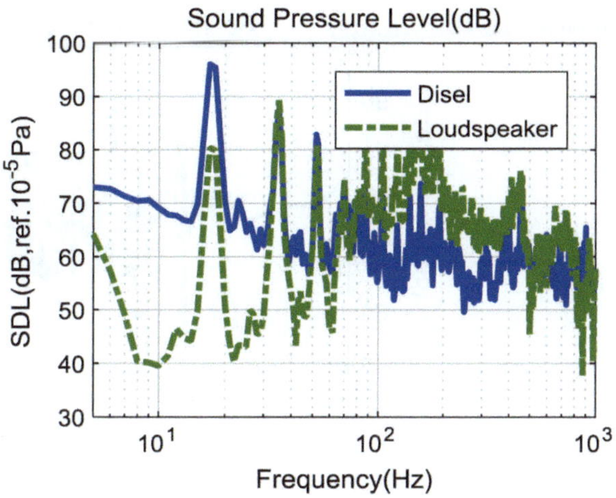

Fig. 5.16 Comparison between sound pressure levels generated by a diesel engine and a loudspeaker collected by a microphone for the idle rotating speed

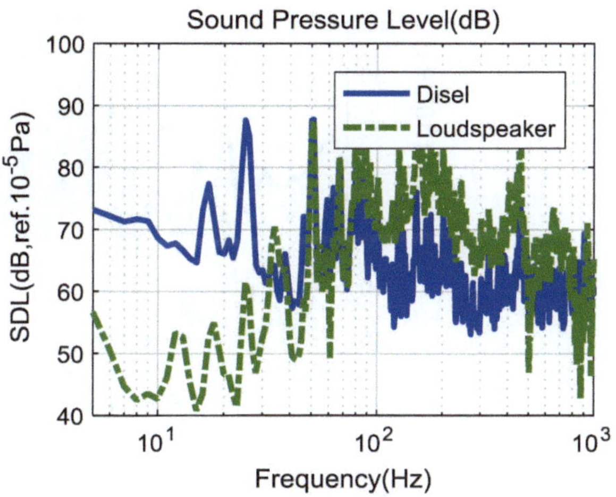

Fig. 5.17 Comparison of sound pressure levels between a diesel engine and a loudspeaker collected by a microphone for the rotating speed of 1000 RPM

average of 4 dB; the effect is not obvious in the frequency band from 20 to 80 Hz; the vibration from 80 to 100 Hz is about 6 dB.

After the active control is applied, the vibration of the diesel engine's foot is greatly amplified in the frequency band below 50 Hz, and the most significant phenomenon is 15 dB for the 15 Hz amplification. The effect on vibration above 50 Hz is not obvious.

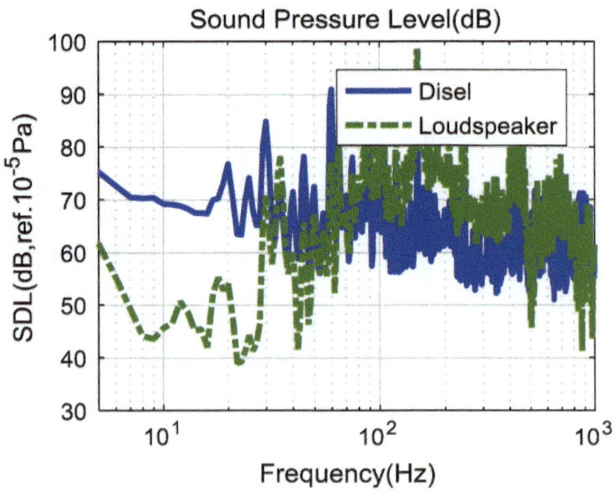

Fig. 5.18 Comparison of sound pressure levels between a diesel engine and a loudspeaker collected by a microphone for the rotating speed of 1200 RPM

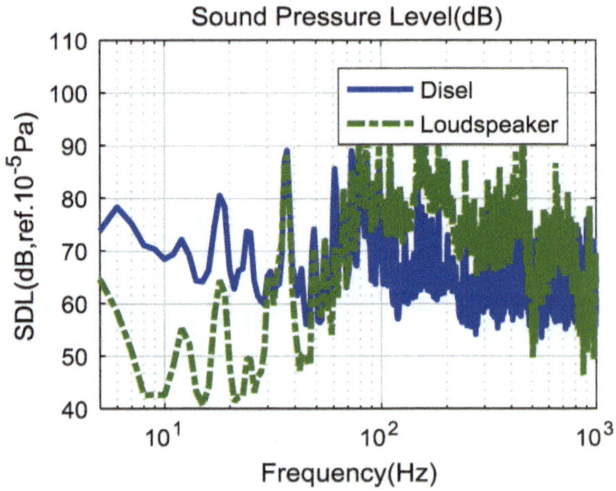

Fig. 5.19 Comparison of sound pressure levels between a diesel engine and a loudspeaker collected by a microphone for the rotating speed of 1500 RPM

For the vibration of the base, after the active control is applied, the vibration in the frequency band below 70 Hz generally decreases, and the average control effect of the vibration protruding line spectrum is 4 dB. In the frequency band from 70 to 100 Hz, the line spectrum vibration is weakly amplified by about 2 dB.

After applying the low-frequency line spectrum adaptive control strategy, for the deck vibration, the vibration amplitude of the vibration prominent line spectrum in the frequency band within 100 Hz has a different amplitude, and the average control effect is 7 dB.

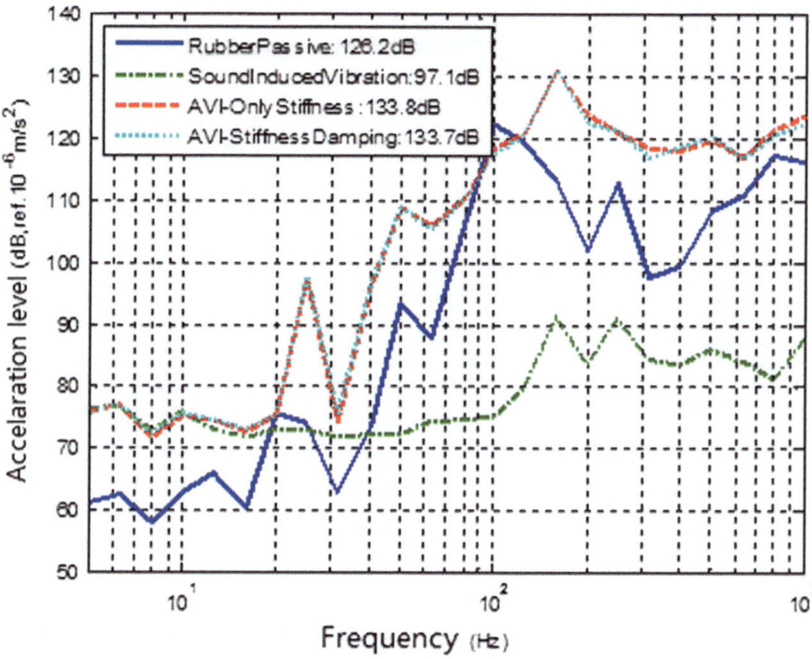

Fig. 5.20 Comparison of vibration levels among various isolation methods at bases for the rotating speed of 1500 RPM

The control of the line spectrum depends on two aspects:

(1) Is this line spectrum prominent in the reference signal?

The more prominent the amplitude of the line spectrum in the reference signal is, that is, the greater the proportion of the total vibration energy, the more attention is paid to the optimal adjustment of the digital filter, and the model parameters for this line spectrum are also the more accurate, the closer it is to the condition represented by Eq. (3.2).

(2) Is the line spectrum prominent in the error signal?

Only the line spectrum corresponding to this frequency is equally prominent in the error signal, i.e., the line spectrum still occupies a higher proportion in the vibration energy at the control target, to show the significant control effect caused by the accurate model. This could be verified from the control effects of frequencies 52, 65, 88, and 95 Hz in Figs. 5.28 and 5.31.

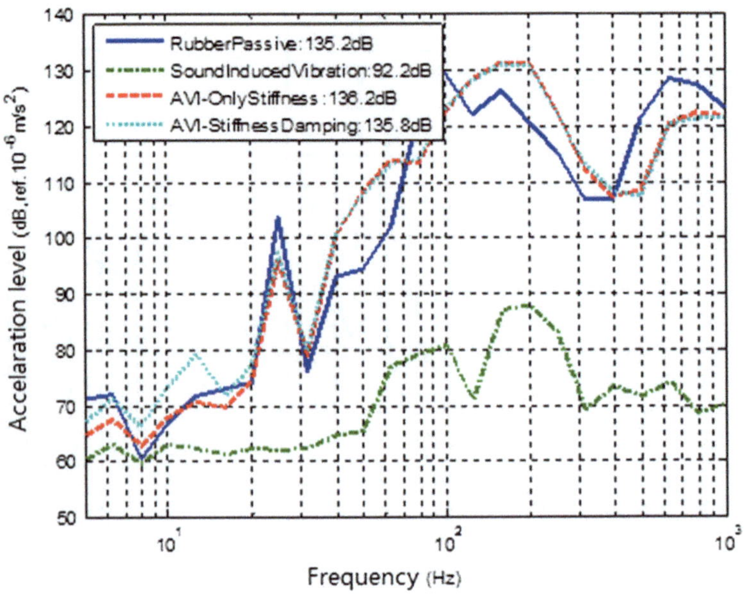

Fig. 5.21 Comparison of lateral vibration levels of diesel engine among various isolation methods for the rotating speed of 1500 RPM

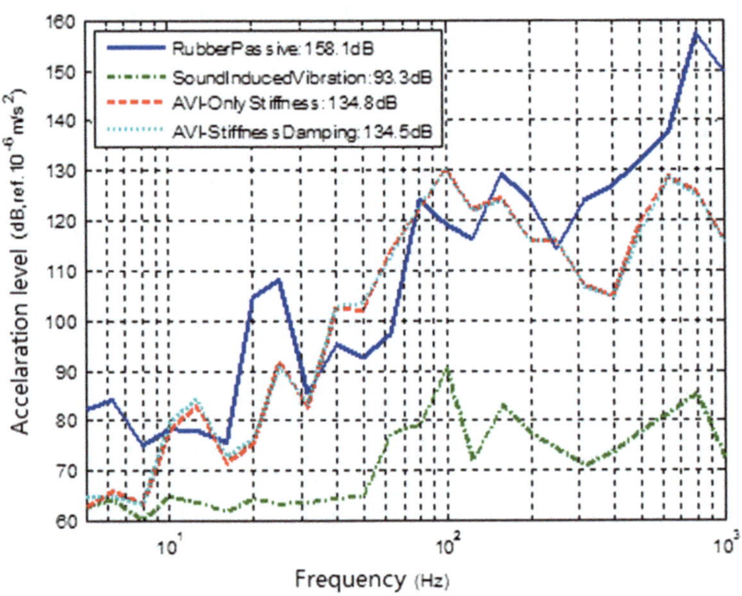

Fig. 5.22 Comparison of longitudinal vibration levels of diesel engine among various isolation methods for the rotating speed of 1500 RPM

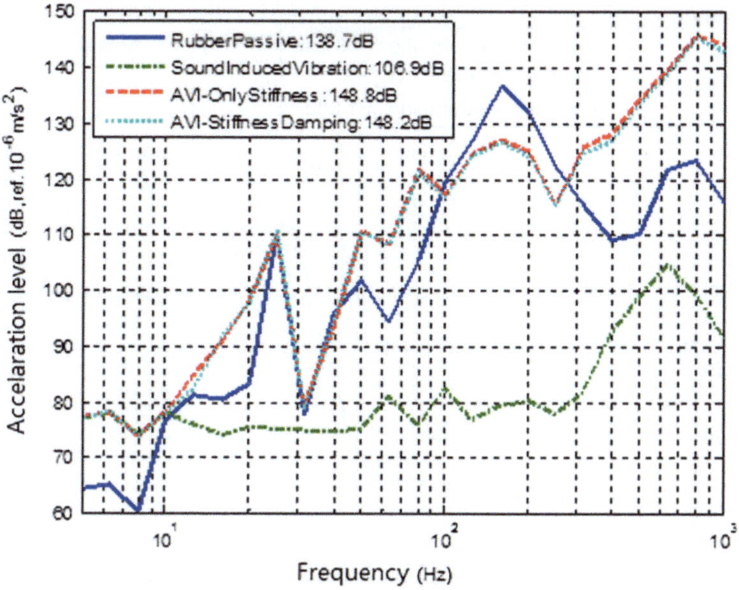

Fig. 5.23 Comparison of vertical vibration levels of diesel engine among various isolation methods for the rotating speed of 1500 RPM

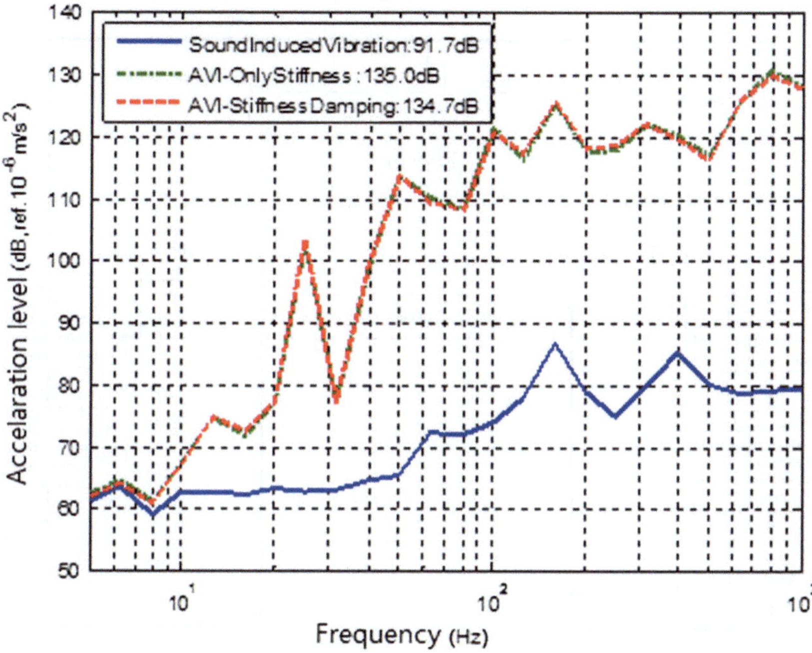

Fig. 5.24 Various vibration levels of feet for the rotating speed of 1500 RPM

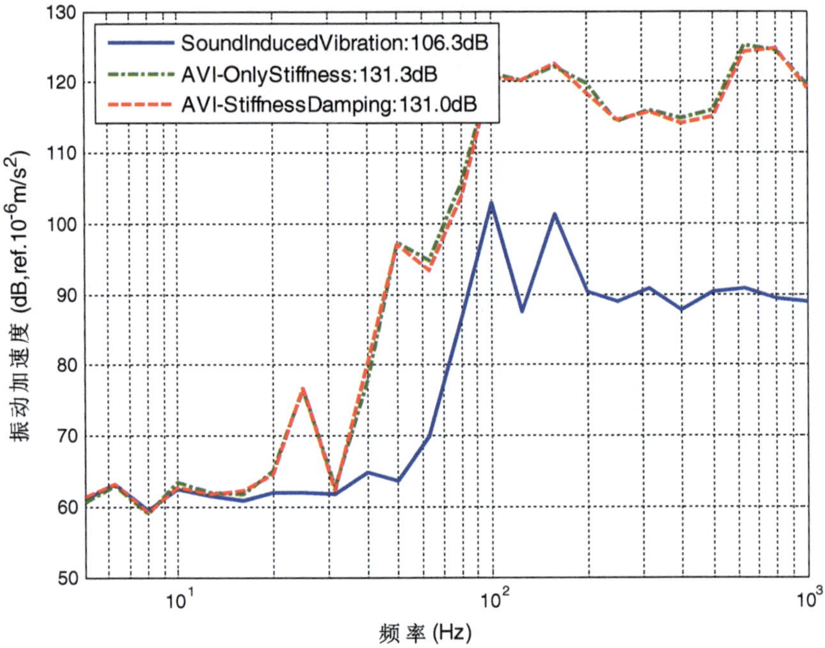

Fig. 5.25 Various vibration levels of deck for the rotating speed of 1500 RPM

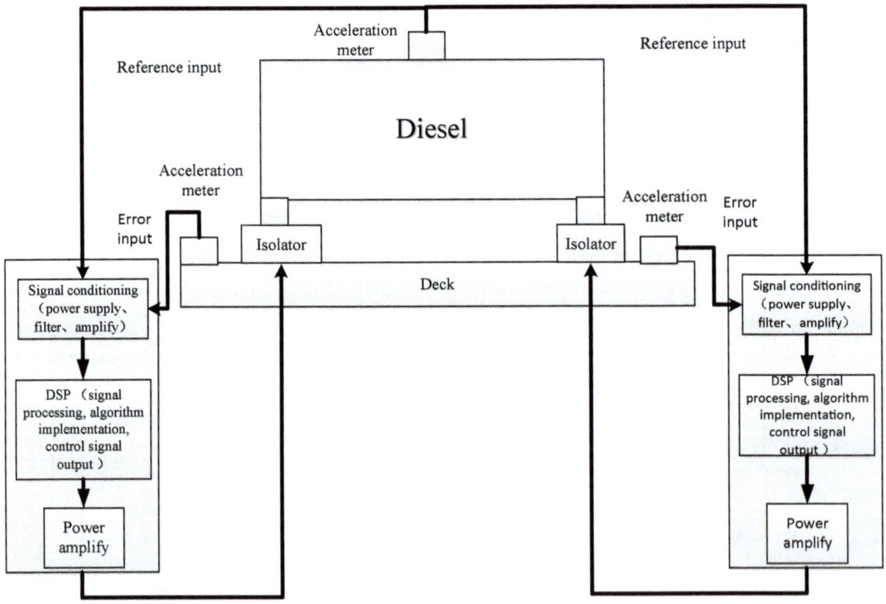

Fig. 5.26 Schematic for active and passive hybrid vibration isolation test

Fig. 5.27 Spot scene of active vibration isolation test

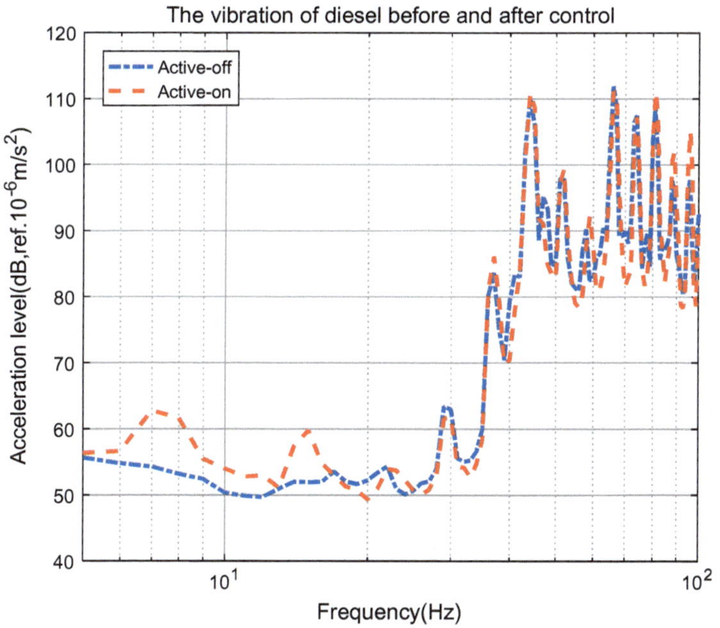

Fig. 5.28 Vibration levels of diesel engine before and after active control for the idle rotating speed

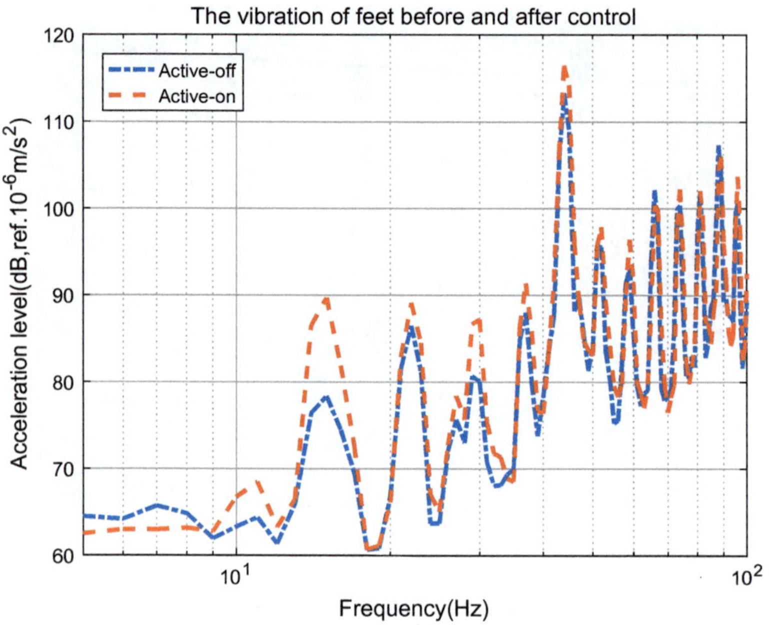

Fig. 5.29 Vibration levels of feet before and after active control for the idle rotating speed

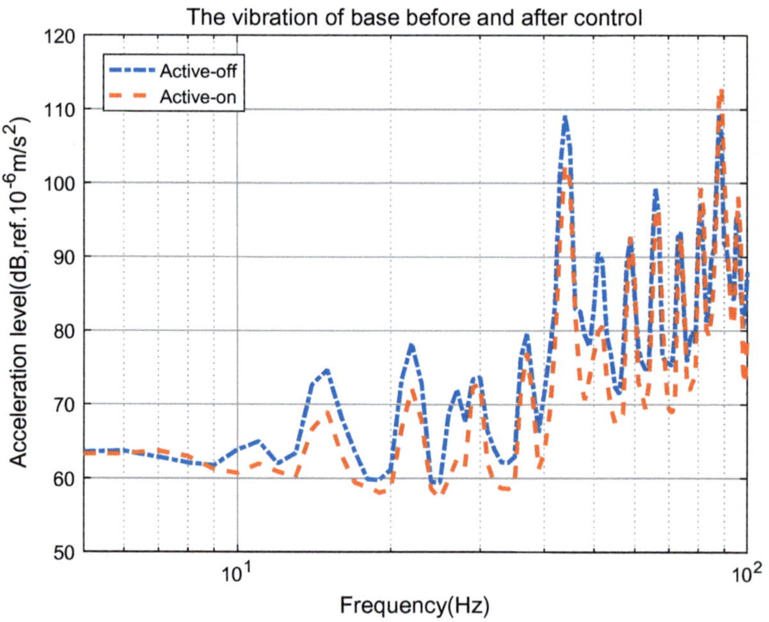

Fig. 5.30 Vibration levels of bases before and after active control for the idle rotating speed

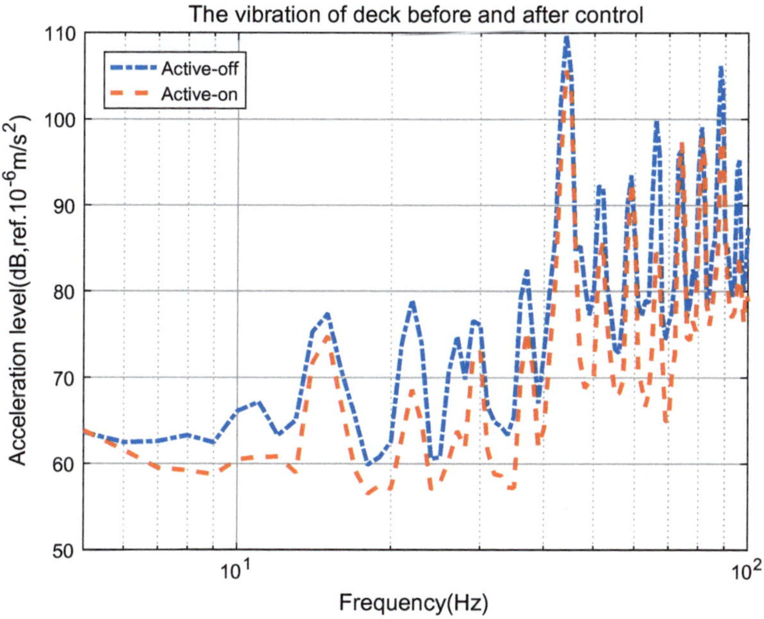

Fig. 5.31 Vibration levels of deck before and after active control for the idle rotating speed

5.5 Conclusion

In this chapter, the vibration isolation effect of the designed active–passive hybrid vibration isolation system is verified under the practical application environment. For the vibration characteristics of the diesel engine in a typical installation environment, the vibrations of the diesel engine's foot, base, and deck were measured under the conditions that the active control vibration isolation unit was not turned on and turned on. Verify the damping effect of active and passive isolators and adaptive feed-forward control. Research shows:

(1) Active control is not enabled. Only passive vibration isolation is used to obtain 10 dB vibration isolation effect for base vibration in the frequency range below 1000 Hz.
(2) Using simulated air noise excitation test measurements to consider the impact of diesel engine noise on the excitation of the base and deck, the active and passive vibration isolation test should be performed in the frequency range above 40 Hz.
(3) Initiate active control and apply the designed control strategy and control algorithm. After applying active control to the base vibration, the average control effect in the frequency range below 70 Hz is 4 dB. For deck vibration, the average control effect in the frequency band below 100 Hz is 7 dB.

Chapter 6
Research on Pipeline Three-Way Adjustable Frequency Dynamic Vibration Absorption Technology

Abstract This chapter studies the three-way adjustable frequency vibration absorption technology of the typical piping system of vessel, proposes the design theory and scheme of the three-way adjustable frequency dynamic vibration absorption, and uses the vibration table to test the vibration performance of the absorber.

6.1 Introduction

The radiation noise caused by the vibration of the piping system of ships transmitted to hull is also an important way to generate mechanical noise. The traditional methods adopted for vibration transmission control of piping system include flexible joints, constrained damping layers, elastic supports, etc. Generally speaking, these methods could hardly control the low frequency vibration of pipelines. In addition, DVA is also an effective way to reduce the pipeline vibration transmission. However, general dynamic vibration absorbers have strong frequency selectivity, and it is not suitable for piping systems with certain frequency bandwidth or frequency variation. With the vibration frequency extending to a low-frequency, electromagnetic, piezoelectric, or magnetostrictive material, actuators are used to active control of the pipeline vibration. Although the low frequency vibration of pipeline could be locally reduced, it is constrained due to cost and power consumption limitations. Therefore, it is necessary to develop a vibration-absorbing system that can adapt to frequency adjustment and vibration absorption according to changes in the piping vibration and reduce power consumption as well as cost substantially.

This chapter combines the vibration absorption characteristics of the vessel piping system, studies the three-way adjustable frequency vibration absorption theory and design method, then designs an adjustable frequency vibration absorber, and finally, a vibrator was used to test the three-way frequency adjustment capability of the designed adaptive dynamic vibration absorber (ADVA).

F. Wang et al., *Comprehensive Investigation on Active-Passive Hybrid Isolation and Tunable Dynamic Vibration Absorption*, Springer Tracts in Mechanical Engineering, https://doi.org/10.1007/978-981-13-3056-8_6

6.2 Vibration Absorption

6.2.1 Passive Dynamic Vibration Absorption

Vibration absorption mainly reduces the vibration of controlled structures through the vibration energy transfer method. Researches on method for the dynamic vibration absorber design include: Cheng [1] proposed to follow the modal times, analyze the vibration absorber parameter effects in order, and design the vibration absorber parameters accordingly, which could provide the best effect of vibration suppression. Researches by Gawthrop et al. [2] show that the dynamically dual vibration absorber (DDVA) method provides a new method for designing shock absorbers in physical field. Jin and Cheng [3] discussed the theory of classical dynamic vibration absorbers and nonlinear dynamic vibration absorbers and then extended it to elastic systems and the possibility of multi-frequency and multi-modal dynamic vibration absorption. Yu et al. [4] adopted a multivariable and multivariable adaptive genetic algorithm to obtain the optimal vibration absorber parameters of the damping system. The mathematical expression of the optimal parameters was obtained by regression analysis, and the optimum design of vibration absorber was optimized through simulation and experiments. Liu and Liu [5] studied the classical problem of an optimally damped dynamic vibration absorber (called model A) and found that the optimal results cannot be obtained using the Den Harto method and the Kelly method. Espíndola et al. [6] studied how to describe a system that uses an objective function defined by the Frobenius norm to design an optimal viscoelastic absorber (known as four fractional parametric models).

Researches on new types of vibration absorbers include: Zeng et al. [7] designed a detachable annular tuned mass damper (Fig. 6.1), established a piping system experimental device, and performed simulation calculations. The vibration of the experimental piping system in time and frequency domain both was analyzed with and without tuned mass damper; meanwhile, the experiment was verified. Results show that the tuned mass damper designed can greatly reduce the steady-state vibration of the piping system and quickly attenuate the transient vibration of the piping system.

Li et al. [8] proposed a design scheme for a dynamic vibration absorber with a compact spatial layout, a non-contact eddy current energy dissipation mechanism, and damping that could be designed for the suppression of aircraft tail buffeting. The test results show that the eddy-flow force absorber has good vibration absorption performance and the maximum reduction ratio can reach 98%. Habib and Kerschen [9] studied the self-excited vibration of a mechanical system using a nonlinear tuned vibration absorber. A significant feature of this type of vibration absorber is that the mathematical model could be customized according to the similar principle of a nonlinear primary system. Benacchio et al. [10] designed a passive dynamic vibration absorber using ring magnets, as is shown in Fig. 6.2, and used the suction and repulsion forces between magnets to change the stiffness of the vibration absorber.

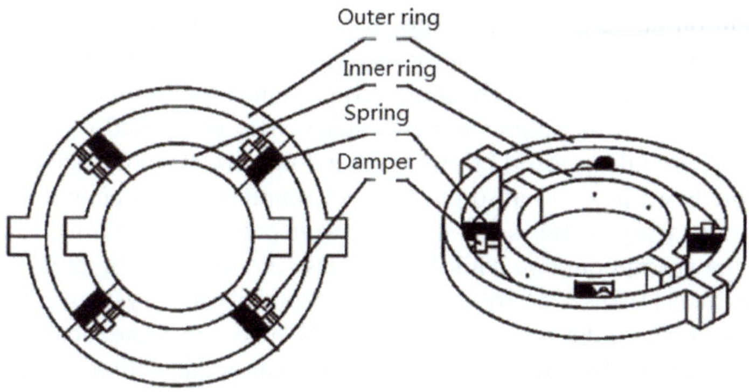

Fig. 6.1 Tuning vibration damper used for pipe

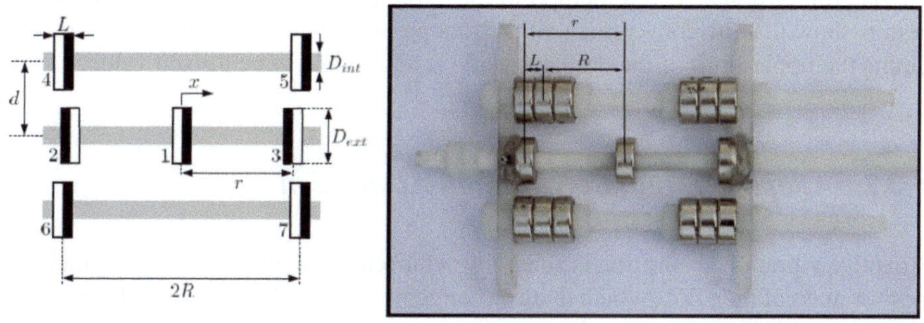

Fig. 6.2 Ring magnet nonlinear dynamic vibration absorber

After adjustment, it could be used as a nonlinear tuned absorber, nonlinear energy trap, and negative stiffness bistable absorber.

Researches on the application of vibration absorbers include: Bonsel et al. [11] applied a linear dynamic vibration absorber to the vibration control of a piecewise linear beam. Two types of undamped and damped dynamic vibration absorbers were studied. Theoretical analysis results show that the undamped dynamic vibration absorber is suitable for the interference frequency constant. What's more, the damped vibration absorber is suitable for the unstable frequency interference, and the above conclusions were obtained through experiments.

Sun et al. [12] analyzed the response characteristics of a single-degree-of-freedom system with two-degree-of-freedom excitation and provided theoretical support for the time span and step-size selection in simulation calculations. The performance of a dynamic shock absorber and a state switching absorber (SSA) was compared. Results show that the double DVA has almost the same performance as the SSA. In addition, compared with SSA, DVA has the advantages of low adjustment frequency, fast optimization response, and low requirement on material fatigue resistance. Webster and Vaicaitis [13] elaborated on the successful application of a tuned mass damping system to reduce the long-span, steady-state vibration of a cantilevered composite floor system on the terrace of the New York City Park Building. Yang et al. [14] discussed the use of dynamic vibration absorbers to control structural vibrations in narrowband and wideband. Kęcik et al. [15] presented the results of a study on the dynamic response of an automatic parameter system composed of a vibrator and an additional pendulum absorber. Analysis and numerical studies have shown that it is possible to achieve full absorption if the damping of the pendulum is close to zero. Damping analysis shows that the increase in pendulum damping can reduce or eliminate the absorption area while the increase in vibration damping reduces the absorption. Yuan [16] used a two-degree-of-freedom vibration system with a dynamic vibration absorber as the research object, established a mathematical model using D'Alembert's principle, and performed a dimensionless calculation. The four parameters (κ, γ, δ, μ) of the dynamic vibration absorber are comprehensively optimized using the Davidon–Fletcher–Powell (DFP) method, the penalty function method, and the one-dimensional search method in the variable metric method. The calculation results show that the convergence of the four parameters is different. γ and δ tend to define the upper limit of the domain, while μ and κ tend to be a fixed value.

6.2.2 *Adaptive Dynamic Vibration Absorption*

Adaptive vibration absorption is generally achieved by adaptively changing the stiffness or mass of the DVA, which in turn changes the stiffness. Li et al. [17] designed a frequency-shifting DVA that adjusts its geometrical parameters such that the natural frequency changes linearly with the geometric parameters, and preliminarily designs the corresponding control method. Xu et al. [18] developed a mechanical self-tuning dynamic vibration absorber that adjusts the natural frequency by adjusting its own geometric parameters. The evaluation experiment conducted on the beam structure experimental platform shows that the natural frequency can change from 14.2 to 47.2 Hz when the span of the DVA changes from 30 to 140 mm, and the frequency change can reach up to 232%, which can satisfy the change of the excitation force frequency. Demand for vibration reduction is in the larger range. The double-mass damper designed by Hill et al. [19] is also a relatively common frequency adjustable damper (Fig. 6.3).

The ADVA designed by Mirsanei et al. [20] uses a servo motor to change the position of the mass on the cantilever beam and adaptively adjusts the natural frequency of the DVA, as is shown in Fig. 6.4.

There are also other design ideas, such as Gong et al. [21] designed a pendulum-type ADVA-based on the pendulum effect and applied it to a multi-modal system. Simulation and test results show that the vibration of the structure could be effectively suppressed over a wide frequency range. Aguirre et al. [22] designed a DVA with self-sensing frequency adjustment for flutter in mechanical systems. By controlling the stiffness of the motor to adjust the spring to adaptively change the resonant frequency of the absorber, the damping is also created by the eddy current phenomenon produced by the vibration of the conductive plate in the magnetic field. Mikułowski

Fig. 6.3 Double-mass DVA

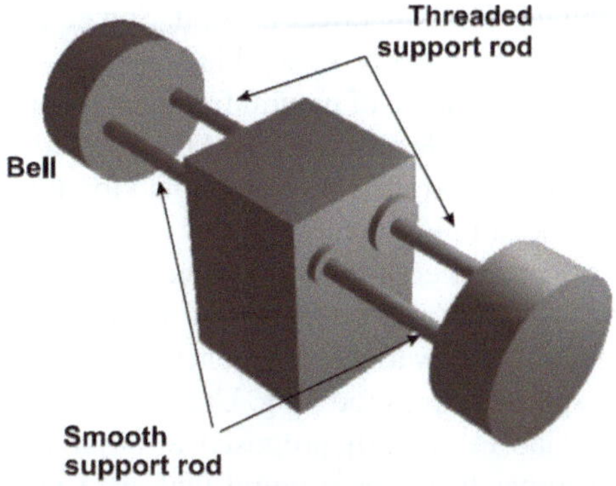

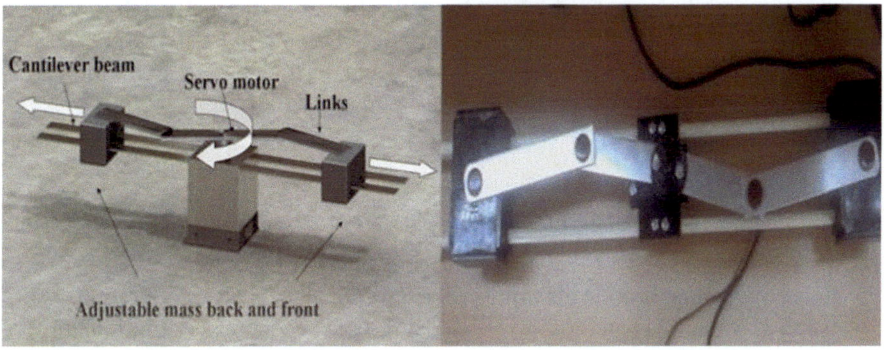

Fig. 6.4 ADVA and slider crank mechanism

and Wiszowaty [23] designed an adaptive airbag shock absorber and verified the effectiveness of the method through experiments.

For nonlinear vibration absorption, Yang et al. [24] designed a nonlinear vibration absorption device suitable for space environment structure. By introducing nonlinear magnetic force to replace the nonlinear spring that is difficult to realize, it was applied to space cantilever. Beam structure.

In addition, with the wide application of new materials, the use of new materials to design ADVA has also become a hot spot. Wang et al. [25] designed a magnetorheological elastomer self-tuning absorber and optimized it with an improved genetic algorithm. Experimental results show that this kind of genetic algorithm has the characteristics of global search and fast convergence. It can make the vibration absorber find the position with the best effect of vibration absorption. The effect of vibration reduction can reach up to 25 dB. Kang et al. [26] designed a controller based on TMS320F2812 DSP as the core to realize vibration damping control of an active self-tuning vibration absorber based on magnetorheological elastomer and proposed a control algorithm combining variable step-size optimization and feedback control. And on the multi-modal experimental platform for vibration absorption vibration absorber experimental study, results show that the optimization time is less than 15 s, and the damping effect is good. Sun et al. [27] designed a magnetorheological fluid elastic shock absorber working in squeeze mode, as is shown in Fig. 6.5. The variable frequency range of natural frequency in compression mode is 37–67 Hz. Herold and Mayer [28] designed a self-ADVA based on piezoelectric materials based on piezoelectric materials and verified its vibration reduction effect through experiments, as is shown in Fig. 6.6.

Rustighi et al. [29] designed an adaptive dynamic vibration absorber (Fig. 6.7) by using the characteristics of memory alloys whose stiffness changes with temperature, with a variable frequency range of 15%. The disadvantage is that the required response time is too long. Using a 9 A current requires 120 s of heating time.

Gao et al. [30] proposed a variable mass tuning adaptive dynamic vibration absorber that uses a liquid tank as a variable mass unit to change the quality by injecting or withdrawing liquid into the tank. To adjust the natural frequency of the absorber, experimental results show that the variable mass dynamic vibration absorber has a wider effective frequency band, and during the test, a vibration attenuation effect of about 29 dB on the main system was achieved.

6.2.3 Active Dynamic Vibration Absorption

For active dynamic vibration absorption, the traditional mechanical vibration absorption, such as Wang [31] and Yan [32], used an electromechanical vibration absorber (Fig. 6.8) with adjustable control force on an experimental bench to carry out an active vibration absorption test. At present, most commonly used are electromagnetic active vibration absorbers. For example, Millitzer et al. [33] designed an active dynamic vibration absorber for the torsional vibration of convertibles (Fig. 6.9) and verified the system by numerical calculations, experimental verifications, and real vehicle tests. With the advancement of technology, some scholars have begun to use new materials to design active dynamic vibration absorbers. As the research of Pagliarulo et al. [34] shows, magnetostrictive DVA could be achieved through appro-

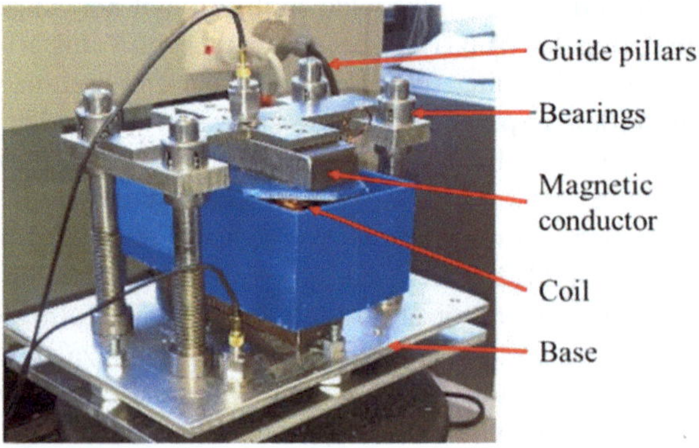

Fig. 6.5 Structure of extruded MRE DVA

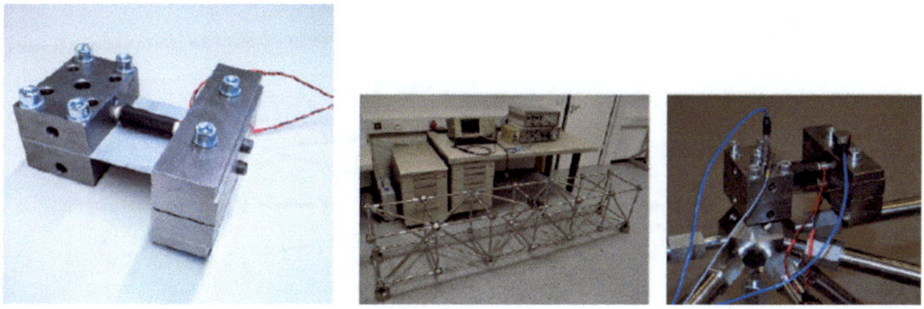

Fig. 6.6 Principle prototype and experimental layout of ADVA

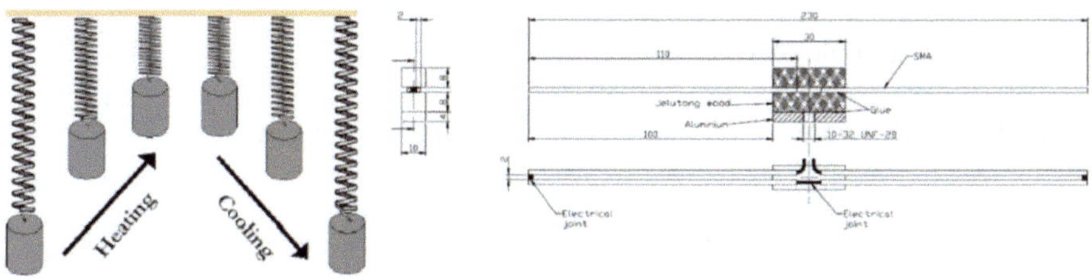

Fig. 6.7 Working principle of shape memory alloy ADVA and design drawing

Fig. 6.8 Adjustable mechanical actuator

priate control law and the configuration of actuators and sensors in the same position (Fig. 6.10). The resonant frequency reduced by about 15%.

Konstanzer et al. [35] designed a piezoelectric vibration absorber as is shown in Fig. 6.11 and applied it to the control of aircraft cabin noise, and control effects of 40 and 35 dB for the fundamental frequencies of the first and second blades were achieved, respectively.

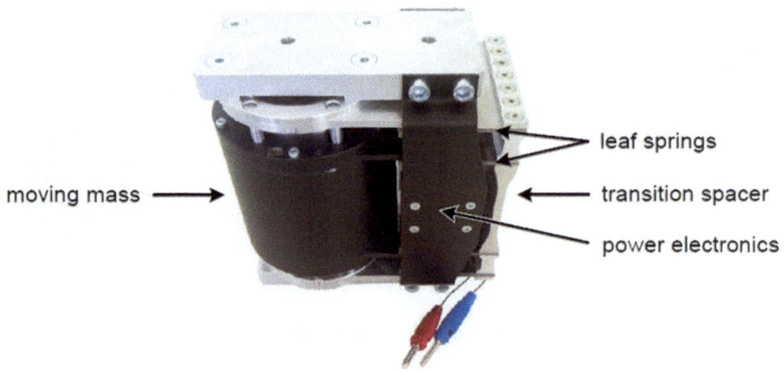

Fig. 6.9 Electromagnetic DVA

Fig. 6.10 Magnetostrictive DVA

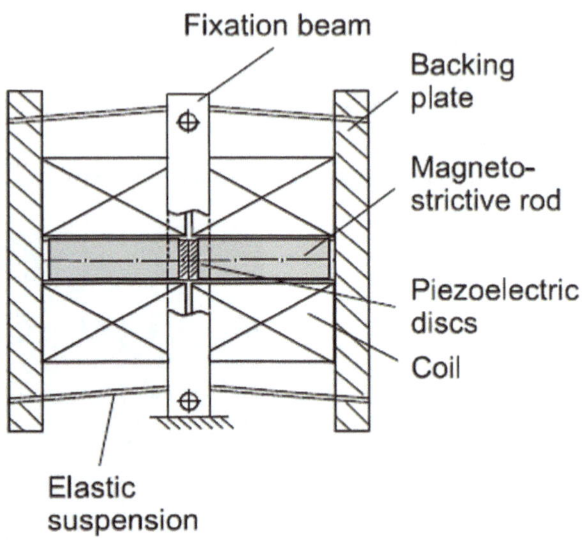

In addition, the best vibration-absorbing effect that could be achieved by active dynamic vibration absorber and the coordination problem between multiple vibration absorbers is also current research hotspots. Chen et al. [36] analyzed the working principle of an active dynamic vibration absorber from the perspective of energy. Results show that the larger the energy transfer coefficient of the active control force, the more obvious vibration control effect could be acquired. Li [37] studied the optimal configuration of actuator installation position in the active vibration control of elastomer and proposed a closed loop configuration optimization criterion based on state feedback and a configuration optimization criterion based on the segmented weighted area performance index. Chen [38] established a floating raft vibration active control experiment system with four active vibration absorbers. To improve the low-frequency vibration-reducing performance of the system, the coordinated

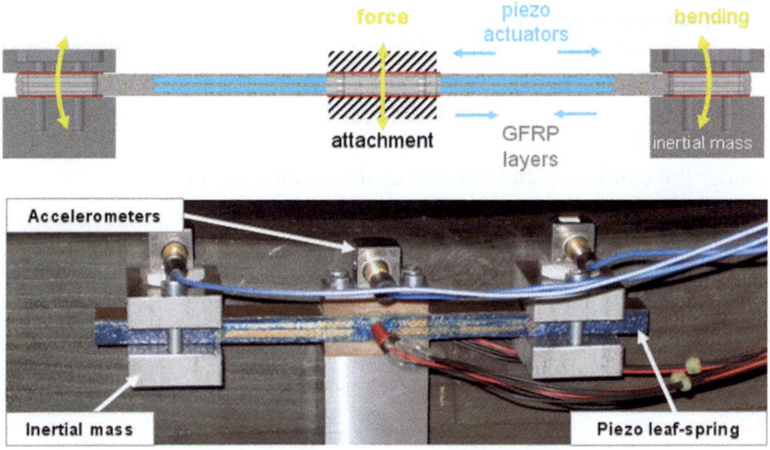

Fig. 6.11 Working principle and prototype of piezoelectric tuning DVA

control of four active vibration absorbers was studied. Experimental results prove the effectiveness of multi-vibrator coordination and active control to improve the floating low-frequency damping capability. Beltran-Carbajal et al. [39] considered the application of a passive–active Duffing absorber in a Duffing mechanical system (possible resonance) under the direct unknown harmonic excitation force.

6.3 Adaptive 3 DOF DVA

6.3.1 DVA Theory for Piping

For passive vibration absorption, there are two main methods for designing of the parameters of DVA:

(1) Impedance coupling method

The impedance coupling method does not require a complete model of the structure. It is only necessary to know the impedance of the drive point where the absorber is installed as well as the impedance of the absorber. In general, the impedance of the vibration absorber needs to be as large as possible in the frequency range of interest, so that the structural response at the mounting position could be reduced by the vibration absorber.

The harmonic force acts on the object could be obtained under steady-state conditions

$$V_p = F_p / Z_p \tag{6.1}$$

where V_p is the speed, and Z_p is the impedance of the driving point.

Similarly, for vibration absorbers there are:

$$V_t = F_t / Z_t \tag{6.2}$$

Consider the continuity and balance of the coupling system.

$$\begin{aligned} V_c &= V_p = V_t \\ F_c &= F_p + F_t \end{aligned} \tag{6.3}$$

Therefore, using Eqs. (6.1) and (6.3), the input and output relationships for the coupling system are as follows:

$$\begin{aligned} V_c &= F_c / Z_c \\ Z_c &= Z_p + Z_t \end{aligned} \tag{6.4}$$

where Z_c is the driving point impedance, and V_c is the speed of the coupling system at the position where the dynamic absorber is installed. Note that the input–output relationship of the coupled system is determined by the characteristics of the independent structure. Equation (6.4) shows that to obtain 20-dB attenuation in the specified frequency response of the structure, the impedance of the coupling system must be 10 times greater than the impedance of the primary system.

(1) Rules of thumb

Assuming that the frequency to be controlled is ω, the relationship between stiffness k_a and mass k_a of the absorber could be obtained as $\omega^2 = k_a / m_a$.

Take the constraints of the output force and movement of the absorber into consideration, select appropriate m_a and k_a and define $\mu = m_a / m$, so that μ satisfy $0.05 < \mu < 0.25$.

Both methods need to know the quality and installation conditions of known mechanical equipment. Based on the two-degree-of-freedom vibration theory, the mechanical device is considered as a rigid body and a dynamic vibration absorber with the best mass ratio or damping ratio is designed. However, when applying the dynamic vibration absorber to the pipeline, considering that the pipeline is a continuous body, the design method based on two degrees of freedom is no longer applicable. Regardless of whether the application object is a mechanical device or a pipeline, the dynamic vibration absorber is essentially a single-degree-of-freedom vibrator attached to the vibrating object. The vibrator has a maximum value of the additional impedance of the vibrator at the resonance frequency point, thereby reducing the object at the resonance frequency point. This additional impedance equals to the result of a series connection of the spring and damper shunt impedance and mass impedance:

$$Z_{\text{Addition}} = \frac{1}{\frac{1}{j\omega M} + \frac{1}{C + K/j\omega}} = \frac{j\omega M (C + K/j\omega)}{j\omega M + C + K/j\omega} \tag{6.5}$$

where M, C, and K are the mass, damping, and stiffness of the shock absorber, respectively, and ω is the corner frequency. Figure 6.12a is a schematic diagram of the use of a dynamic vibration absorber in the pipeline, assuming 50 Hz is the resonant frequency of the absorber. For a dynamic damper with the construction mass $M = 10$ kg, damping $C = 200$ N s/m, the stiffness $K = (2\pi \times 50)^2 \times M = 9.87 \times 105$ N/m, the impedance curve is shown in Fig. 6.12b. At 50 Hz, the additional impedance of the absorber to the pipe reaches an extreme value, thus damping the vibration effect.

One of the problems in designing a pipe dynamic vibration absorber using the above-mentioned additional impedance method is how to select the mass of the vibration absorber. Due to the complex structure of the piping system of ships, it is difficult to design DVA for the dynamic characteristics of the whole piping system. On the other hand, the design of a DVA is also less versatile. Therefore, the DVA applied to the pipeline should be designed according to the section size of the pipeline meanwhile considering the actual space structure of the pipeline.

The pipeline is abstracted as a beam model, and the apparent DVA is abstracted as a discontinuous point on the pipeline. The influence of the discontinuity on the infinitely long beam with equal section is studied. Taking the discontinuity point as the origin of the coordinate and the axis of the beam as the x-axis, a coordinate system as is shown in Fig. 6.13 is established, where the additional impedance of the discontinuity point to the beam could be represented by three impedance components Z_x, Z_y and Z_{xy}. When an axial wave propagates along the coordinate axis in a forward position passing through the discontinuity of the origin of the coordinate, a reflected wave and a transmitted wave are generated. On the left side of the origin, the incident wave and the reflected wave are superimposed to form the following displacement field.

$$u_1(x, t) = A_a e^{j(\omega t - k_a x)} + B_a e^{j(\omega t + k_a x)} \tag{6.6}$$

The displacement field to the right of the origin of coordinates is as follows:

$$u_2(x, t) = C_a e^{j(\omega t - k_a x)} \tag{6.7}$$

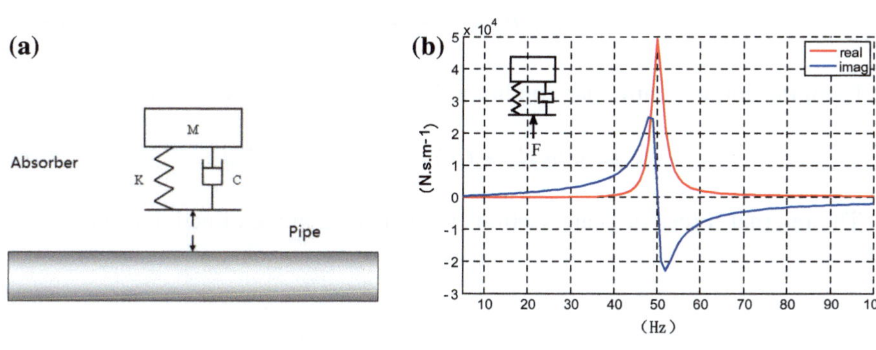

Fig. 6.12 Impedance characteristics of a single-degree-of-freedom oscillator in pipeline

Fig. 6.13 Theoretical model of pipeline DVA

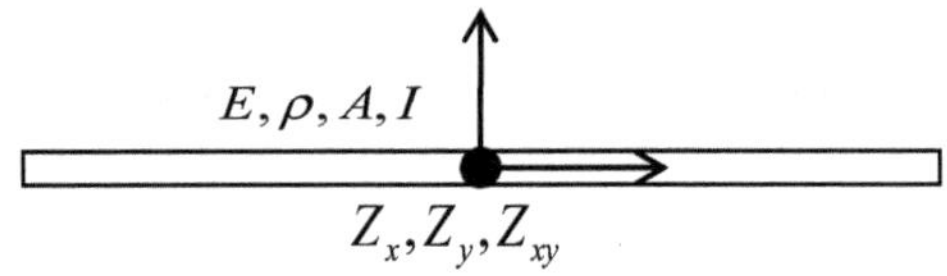

According to the continuous conditions of displacement of the two displacement fields at the origin of the coordinates $u_1(0, t) = u_2(0, t)$,

$$A_a + B_a = C_a \tag{6.8}$$

At the origin of the coordinate, the internal force balance of the beam is as follows:

$$\begin{aligned} & -EA\frac{\partial u_1(0,t)}{\partial x} - Z_x\frac{\partial u_1(0,t)}{\partial t} + EA\frac{\partial u_2(0,t)}{\partial x} \\ & = -EAjk_a(-A_a + B_a) - Z_xj\omega(A_a + B_a) - EAC_ajk_a = 0 \end{aligned} \tag{6.9}$$

According to the above two formulas, the reflection coefficient of the axial wave could be obtained.

$$r_1 = \frac{B_a}{A_a} = -\frac{Z_x}{2EAk_a/\omega + Z_x} \tag{6.10}$$

If the bending wave propagates along the beam, the lateral displacements on the left and right sides of the discontinuity are as follows:

$$w_1(x, t) = A_b\mathrm{e}^{j(\omega t - k_b x)} + B_b\mathrm{e}^{j(\omega t + k_b x)} + C_b\mathrm{e}^{j\omega t + k_b x} \tag{6.11}$$

$$w_2(x, t) = D_b\mathrm{e}^{j(\omega t - k_b x)} + E_b\mathrm{e}^{j\omega t - k_b x} \tag{6.12}$$

From displacement continuous conditions $w_1(0, t) = w_2(0, t)$, we can get

$$A_b + B_b + C_b = D_b + E_b \tag{6.13}$$

From the corner continuous condition $\frac{\partial w_1(0,t)}{\partial x} = \frac{\partial w_2(0,t)}{\partial x}$, we can get

$$-A_bjk_b + B_bjk_b + C_bk_b = -D_bjk_b - E_bk_b \tag{6.14}$$

The internal forces of beam sections are balanced in the direction of the y-axis, i.e.,

$$
\begin{aligned}
& -EI\frac{\partial^3 w_1(0,t)}{\partial x^3} - Z_y\frac{\partial w_1(0,t)}{\partial t} + EI\frac{\partial^3 w_2(0,t)}{\partial x^3} \\
& = -EIk_b^3(A_b j - B_b j + C_b)\mathrm{e}^{j\omega t} - Z_y j\omega(A_b + B_b + C_b)\mathrm{e}^{j\omega t} \\
& + EIk_b^3(D_b j - E_b)\mathrm{e}^{j\omega t} = 0
\end{aligned} \tag{6.15}
$$

Discontinuity points are balanced within the plane, i.e.,

$$
\begin{aligned}
& -EI\frac{\partial^2 w_1(0,t)}{\partial x^2} - Z_{xy}\frac{\partial^2 w_1(0,t)}{\partial x \partial t} + EI\frac{\partial^2 w_2(0,t)}{\partial x^2} \\
& = -EIk_b^2(-A_b j - B_b j + C_b)\mathrm{e}^{j\omega t} - Z_{xy} j\omega k_b(-A_b j + B_b j + C_b)\mathrm{e}^{j\omega t} \\
& + EIk_b^2(-D_b j + E_b)\mathrm{e}^{j\omega t} = 0
\end{aligned} \tag{6.16}
$$

According to Eqs. (6.13) and (6.14), we can get $D_b = A_b + B_b j + C_b(1 + j)$ and $E_b = B_b(1 - j) - C_b j$, substituting it into Eqs. (6.15) and (6.16).

$$
\begin{aligned}
2EIk_b^3(B_b + C_b)(1 - j) &= -Z_y j\omega(A_b + B_b + C_b) \\
2EIk_b(B_b - C_b j) &= -Z_{xy} j\omega(A_b j - B_b j - C_b)
\end{aligned} \tag{6.17}
$$

Divide the two sides of the formula group by A_b, and the formula group obtained after sorting could be expressed as a matrix.

$$
\begin{bmatrix} 1 & 1 \\ 1 & -j \end{bmatrix} \begin{Bmatrix} B_b/A_b \\ C_b/A_b \end{Bmatrix} = \begin{Bmatrix} b_1 \\ b_2 \end{Bmatrix} \tag{6.18}
$$

Solving

$$
r_2 = \frac{B_b}{A_b} = \frac{1-j}{2}(b_1 j + b_2) \tag{6.19}
$$

$$
r_3 = \frac{C_b}{A_b} = \frac{1-j}{2}(b_1 - b_2) \tag{6.20}
$$

$$
r_4 = \frac{D_b}{A_b} = 1 + r_2 j + r_3(1 + j) = 1 + \frac{1-j}{2}(b_1 j - b_2) \tag{6.21}
$$

Among them, $b_1 = -\frac{Z_y}{2EIk_b^3(1-j)/j\omega + Z_y}$ and $b_2 = \frac{Z_{xy}}{2EIk_b/\omega + Z_{xy}}$, r_2 and r_4 are the reflection and transmission coefficients of bending waves, respectively.

According to Eqs. (6.10) and (6.19), it could be seen that the larger the impedance of the additional impedance relative to the pipeline, the larger the reflection coefficient. To increase the additional impedance, the mass of the vibration absorber must also be increased, so that the volume will also increase, which on the one hand increasing the pipeline load, and on the other hand is not conducive to installation. Therefore, it is quite benefit to take the additional impedance and the pipeline impedance into account. In this case, the inertial force of the vibration absorber is equivalent to the dynamic strength level of the pipeline, which can suppress the vibration of the pipeline.

6.3.2 Design Method for Adaptive 3 DOF DVA

In view of the existence of three-way vibration in the pipeline and its mutual conversion during the transmission process, the DVA is used to control the vibration of the pipeline, and the DVA should be able to simultaneously control the vibration in three directions of the pipeline. There are many ways to install a single-degree-of-freedom vibrator on the pipeline. In this case, the cantilever beam is used as the elastic element, and two sets of cantilever beams share the same mass to achieve three-way vibration absorption. As is shown in Fig. 6.14, the two sets of cantilever beams are rigidly connected to the pipe and have a 45° angle with the coordinate axes. The stiffness is provided laterally x and y in two directions. The two sets of cantilever beams work together to provide the stiffness in the z axial direction, so that the natural frequency of z-direction is as follows:

$$f_z = \sqrt{f_x^2 + f_y^2} \tag{6.22}$$

According to the introduction of the research status of semi-active vibration absorption in the first chapter, we can see that there are currently many ways to change the natural frequency of the vibration absorber, and through comparative analysis of various implementation methods, combined with stability, reliability, maintainability, meanwhile considering the limitation of size and weight, it was decided to change the natural frequency of the absorber by changing the position of the slider mass on the cantilever beam and then changing the effective stiffness of the absorber. Using a stepper motor as the actuator, the screw is used to convert rotational motion to linear motion and dragging the slider to advance or retreat, as a result changing the effective stiffness of the absorber, as is shown in Fig. 6.15.

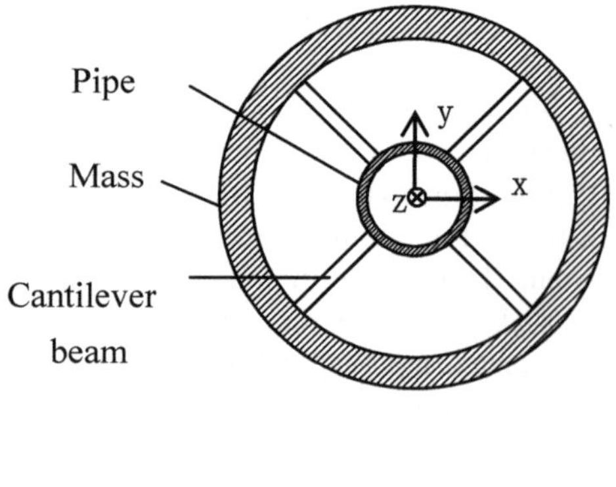

Fig. 6.14 Pipeline three-way cantilever beam dynamic vibration absorber schematic

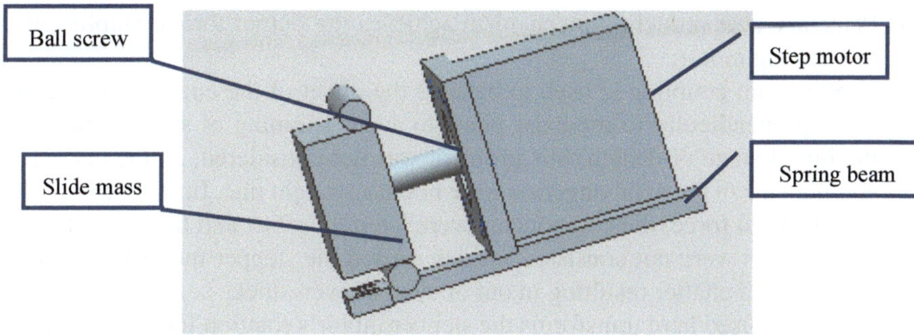

Fig. 6.15 Schematic of executive structure

In addition, due to the size limitation, even if the maximum stepper motor space that allowed is used, the slider cannot be moved while the spring deformed. Therefore, it is necessary to increase a torque amplifying mechanism, so the planetary deceleration mechanism is introduced here, which could achieve deceleration and torque amplification, as is shown in Fig. 6.16.

where

(1) Stepper motor 42 is derived by adjusting the rotation direction and the number of rotation steps under control, and the self-adapting frequency of DVA changes adaptively according to working conditions.

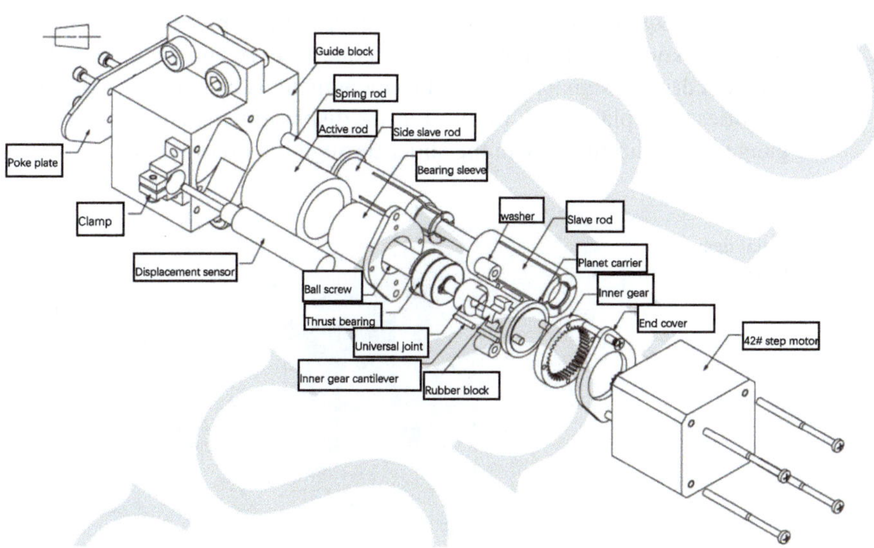

Fig. 6.16 Details of implementation structure

(2) Planetary gear reduction mechanism achieves the output torque amplification of stepping motor;
(3) The Oldham coupling is used to balance the offset of the entire structure in a plane perpendicular to the axial plane to avoid jamming of stepper motor. In the initial stage of design, this problem was not considered, so the centers of components in the axial direction were not in a straight line. In other words, due to the lateral force, the components were laterally offset and the centers of the components were not consistent, which caused the stepper motor to be forced laterally and chatter resulting in out of sync or even stuck.
(4) The screw used here transforms the stepper motor's rotation into linear motion to move the slider forward and backward. On the other hand, it further enlarges the output torque of step motor and locks the position of the slider preventing the unwanted movement of system. Due to the vibration of piping system, the position change of slider would change the natural frequency of DVA and reduce the absorption effect.
(5) The bearing sleeve belongs to the structural design of the piston and the cylinder, which is used to reduce the sliding friction and provide a good guide for forward and backward movement of the structure, avoiding the stalling of the slider in forward and backward movement due to the forward and backward inclinations. The initial design scheme is shown in Fig. 6.17, and it was found in the test that due to the excessively small guide surface; the friction of the slider may increase due to forward and backward tilt during operation.

    (1) The displacement meter used here is to accurately feedback the position of the slider on the spring rod on line, which on the one hand to achieve accurate frequency regulation, on the other hand to prevent the controller from not knowing that the slider has gone to both ends and continue to send out driving signals. The displacement meter is installed by a jig which facilities the replacement. In addition, the displacement of the displacement gauge is synchronized with the slider through the tip of the dial.

Fig. 6.17 Initial design prototype of the slider mass

(2) The spring rod provides the stiffness of DVA. By changing the position of the slider on the spring rod, the effective stiffness of the spring could be changed, and then the natural frequency of ADVA could be adjusted.

It should be noted that when vibration is absorbed, that is, after the controller moves the slider to the specified position of the spring rod, the spring rod should be in close contact with the slider. Only then can the stiffness of the spring rod indicate the effective stiffness of the ADVA in this direction. In addition, it is required that during the movement of the slider, the slider is preferably separated from the spring rod to reduce friction. Thus, the design focusses on how to achieve the above two requirements. The initial design *i* of the spring plate is shown in Fig. 6.17. The assembly test shows that the sliding friction is too large; especially when the spring piece deformed due to the mass of the absorber, the friction force rises sharply and the step motor cannot drive the slider to move. The revision schematic includes two aspects:

(1) Changing the spring piece into a spring rod has the advantage of reducing the sliding friction and providing stiffness in both directions. That is, the stiffness could be combined in two directions to synthesize the stiffness in the third direction, reducing one direction, i.e., reduces the size, weight, and cost of the DVA.
(2) The slider does not contact with the spring rod directly but through a follower mechanism to achieve close contact. The follower mechanism is divided into two forms of internal and external openings, through the deformation of its own restoring force to form a static close contact with the spring rod. When moving the inner slave mechanism, they would out of contact so that reduce sliding friction.

In addition, for ease of installation, the vibration absorber was designed into two parts, so that the installation of the vibration absorber could be achieved without disassembling the pipeline. In addition, the vibration absorber was designed to have inside and outside two parts. The internal is in the form of a ring clamp with a fastening and adjustment mechanism. Therefore, even if the piping is slightly deformed, it can also be tightly installed. The external part is the electromechanical functional structure. The inside and outside are connected through spring rods. The overall installation diagram is shown in Fig. 6.18.

6.4 Test of Frequency Adjustment Abilities

Before performing the vibration absorption performance test, it is necessary to know the frequency adjustment performance of ADVA clearly, so that control parameters in controller could be set accurately. The test is performed on the V9mkII electrodynamic vibration table (Fig. 6.19). Performance indexes of the vibration table are shown in Tables 6.1, 6.2, and 6.3.

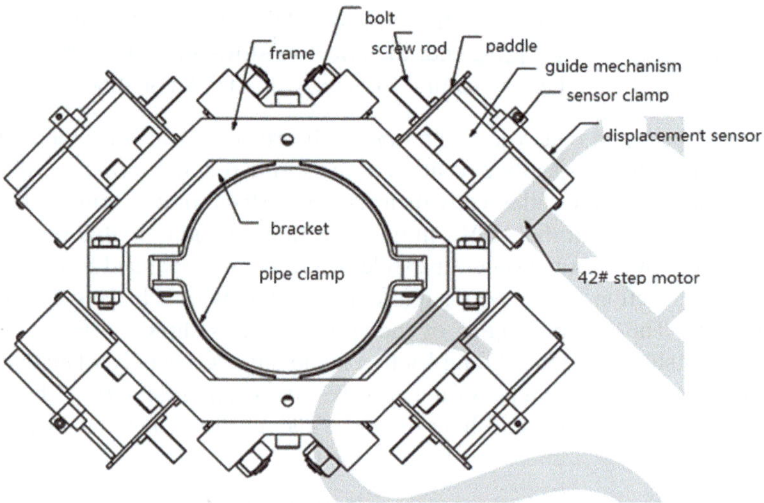

Fig. 6.18 Overall structure diagram of three-degree-of-freedom ADVA

Fig. 6.19 V9mkII electric shaker

The ADVA was mounted on the vibration table, as is shown in Fig. 6.20. Under this condition, the frequency sweep test could be performed in *X* and *Y* directions. After completing the frequency sweep test of one direction, rotate the ADVA by 90° and adjust the mounting positions of sensors accordingly. In addition, to determine the minimum and maximum adjustable natural frequencies that could be achieved in each direction, the slider must be adjusted to the extreme positions of both ends during each test. Using sine signal excitation, the sweep frequency range is set to 20–100 Hz and the acceleration is 1 g.

Firstly, the *X*-direction frequency adjustment performance test is performed. Positions of two sliders corresponding to the excitation direction are adjusted to minimize

Table 6.1 V9mkII shaker table specifications

Parameter	Value
The diameter of working table	432 mm
Vertical extension table	1200 mm × 1200 mm
Sine peak thrust	90 kN
Random RMS lift	105 kN
Semi-sine peak impact thrust	315 kN
Frequency range in use	1.0–2700 Hz
Mass of moving parts	49.8 kg
Sine peak speed	3 m/s
Sinusoidal peak acceleration	1470 m/s^2 (150g_n)
Random RMS acceleration	686 m/s^2 (70g_n)
Maximum (continuous) peak-to-peak displacement	76 mm
Carrying capacity	1800 kg

Table 6.2 Water smoothing indicator

Parameter	Value
The dimension of slide	1220 mm × 1220 mm
System quality (dynamic circle, driving cow heads, and skateboards)	234.5 kg

Table 6.3 Laser vibration controller specifications

The dynamic range of close loop control	Random > 90 dB Sine − 100 dB
Function	Random Sweep sine Sine plus random Random plus random RSTD resonance search and resident Classic impulse TTH transient response control SRS shock response spectrum

the natural frequency in X-direction. The result of the frequency sweep is shown in Fig. 6.21.

Adjust the position of the two sliders corresponding to the excitation direction to maximize the natural frequency in this direction. Results of sweep are shown in Fig. 6.22.

From Figs. 6.21 and 6.22, the adjustable frequency range in the X-direction is 33.2–76.4 Hz.

Fig. 6.20 Test of horizontal excitation

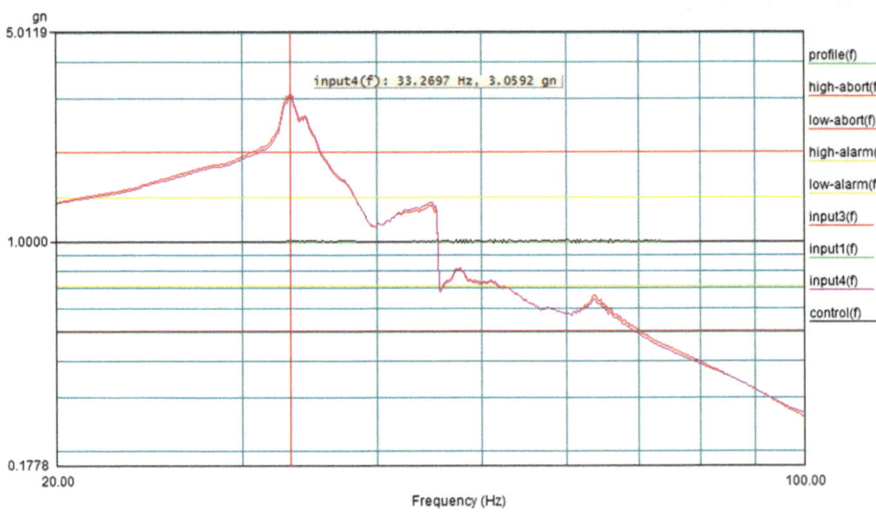

Fig. 6.21 Minimum frequency that X-direction could achieve

Then perform frequency adjustment performance test in Y-direction and adjust the positions of the two sliders corresponding to the excitation direction to minimize the natural frequency in the direction. Results of frequency sweep are shown in Fig. 6.23.

Adjust the positions of the two sliders corresponding to the excitation direction to maximize the natural frequency in this direction. Results of sweep are shown in Fig. 6.24.

From Figs. 6.23 and 6.24, the adjustable frequency range in Y-direction is 37.8–54.5 Hz.

Finally, the Z-direction frequency adjustment performance test was performed, and the arrangements of sensors are shown in Fig. 6.25.

Adjust the positions of four sliders to minimize the natural frequencies both in X and Y directions. Results of sweep are shown in Fig. 6.26.

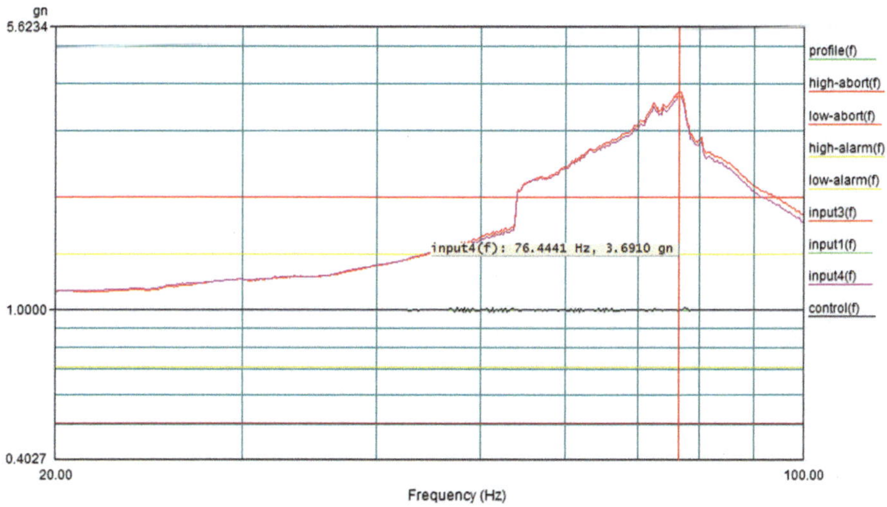

Fig. 6.22 Maximum frequency that X-direction could achieve

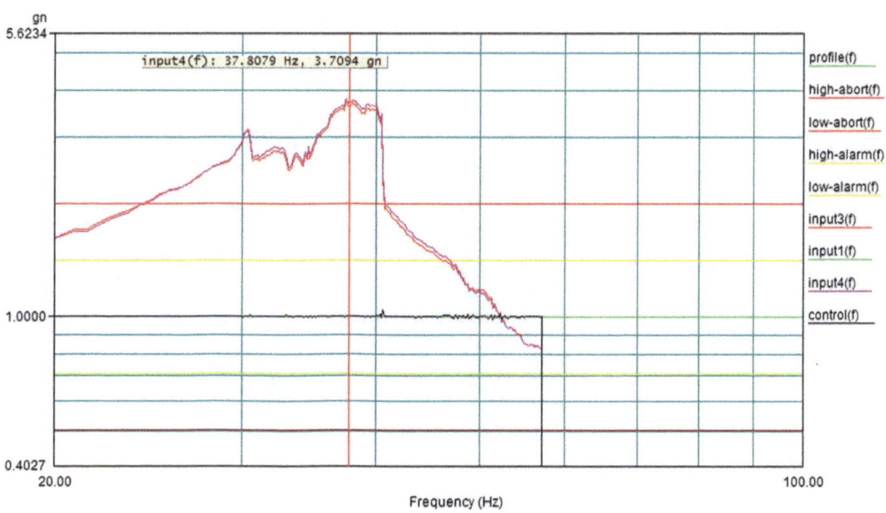

Fig. 6.23 Minimum frequency that Y-direction could achieve

Adjust the positions of the four sliders to maximize the natural frequencies both in X and Y directions. Results of sweep are shown in Fig. 6.27.

From Figs. 6.26 and 6.27, the adjustable frequency range in Z-direction is 44.8–93.7 Hz.

Theoretically, the adjustable frequency in X-direction and Y-direction should be the same, but due to the presence of manufacturing errors, the start and end frequencies and the adjustable range of the two are not the same. Specifically, the maximum

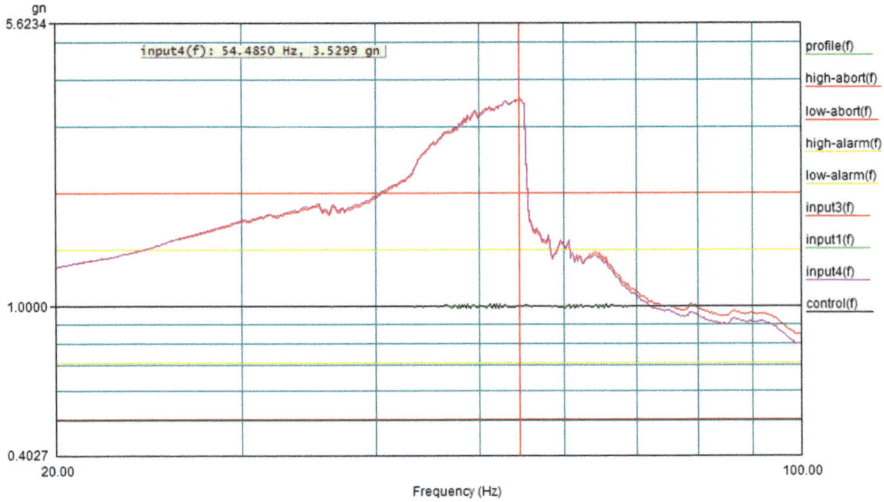

Fig. 6.24 Maximum frequency that Y-direction could achieve

Fig. 6.25 Test of vertical excitation

frequency in Y-direction is lower than the maximum frequency in X-direction. At the same time, by observing Fig. 6.24, it could be seen that it falls rapidly at the maximum reachable frequency, because the gap between the movable slider and the spring rod is larger than the amplitude corresponding to the maximum frequency, which result in a rapid decrease in the resonant peak due to the insufficient processing accuracy. Normal conditions should be that shown in Fig. 6.22.

In addition, according to the design principle (5.22), we can see as follows:

The theoretical minimum frequency in Z-direction is $\sqrt{33.2^2 + 37.8^2} = 50.3$, and the actual test frequency is 44.8 Hz.

The theoretical maximum frequency in Z-direction is $\sqrt{76.4^2 + 54.5^2} = 93.8$, and the actual test frequency is 93.7 Hz, which verify the correctness of the design.

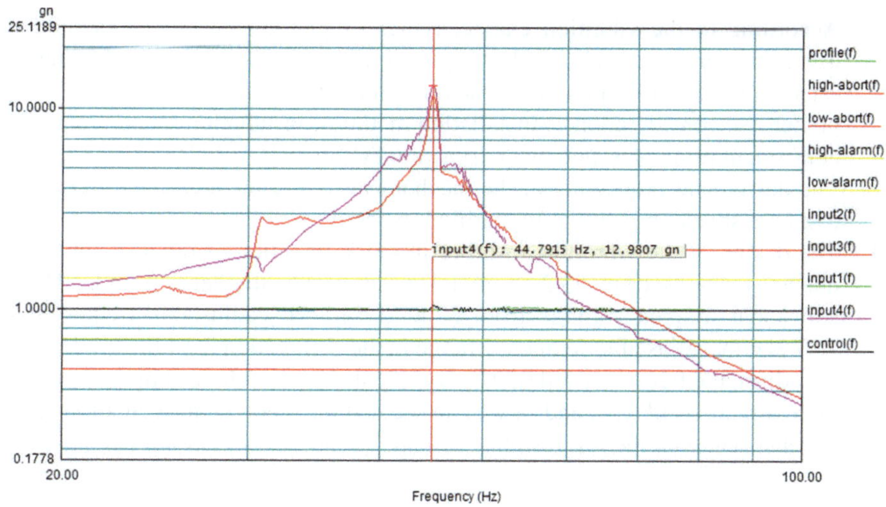

Fig. 6.26 Minimum frequency that Z-direction could achieve

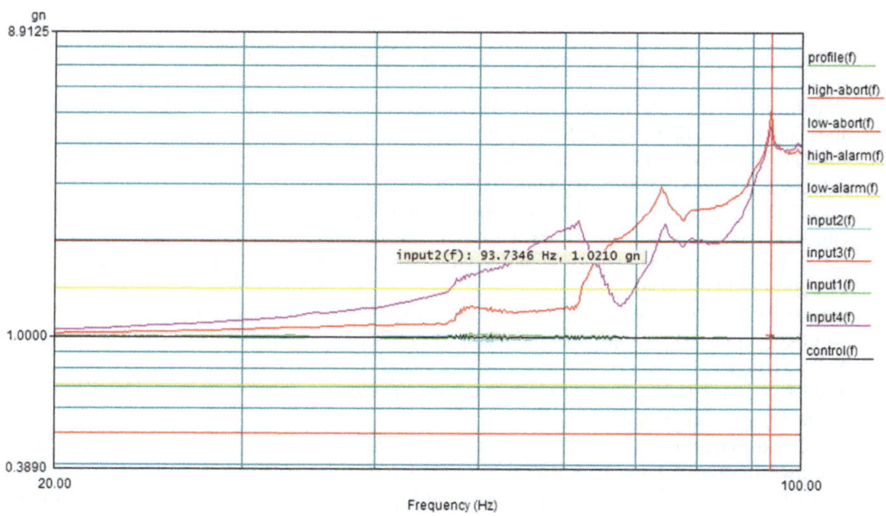

Fig. 6.27 Maximum frequency that Z-direction could achieve

6.5 Conclusion

This chapter combines the characteristics of vibration reduction of the piping system of vessels, studies the theory and design method of three-way adjustable frequency vibration absorption, and designs an adjustable frequency vibration absorber electromechanical actuator. Finally, the three-way frequency adjustment capability of the absorber was tested and verified using a vibration table. Research shows as follows:

(1) According to the vibration characteristics of piping system, the method of designing three-way adjustable frequency dynamic dampers for pipelines is established based on pipeline dynamic vibration absorption theory and comprehensive consideration of installation methods, installation space, frequency adjustment principles, and other constrain factors.
(2) The design and processing of the mechanical part of the adjustable frequency dynamic shock absorber were clarified in detail.
(3) The design of the vibration absorber in the X-direction of the adjustable frequency range is from 33.2 to 76.4 Hz, adjustable frequency is 43.2 Hz; Y adjustable frequency range from 37.8 to 54.5 Hz, adjustable frequency is 16.7 Hz; the adjustable frequency range in the Z-direction is 44.8–93.7 Hz, and the adjustable frequency is 48.9 Hz. By calculating the relationship between the three directions of X, Y, and Z, we can see that the theoretical calculation result is consistent with the test result.

References

1. Zheng RH (2012) Design and analysis of discrete absorbers for beam-type structures. J Technol 2(27):81–90
2. Gawthrop P, Neild SA, Wagg DJ (2015) Dynamically dual vibration absorbers: a bond graph approach to vibration control. Syst Sci Control Eng 3(1):113–128
3. Jin XD, Cheng XM (1997) Ship's ankle vibration and dynamic vibration absorber vibration reduction. J Shang Hai Jiaotong Univ 2:38–40
4. Yu ZF, Wang T, Shen HJ et al (2013) Application of dynamic vibration absorber in flywheel vibration control. Noise Vibr Control 33(5):173–178
5. Liu K, Liu J (2005) The damped dynamic vibration absorbers: revisited and new result. J Sound Vibr 284(3):1181–1189
6. Espíndola JJ, Pereira P, Bavastri CA et al (2009) Design of optimum system of viscoelastic vibration absorbers with a Frobenius norm objective function. J Braz Soc Mech Sci Eng 31(3):210–219
7. Zeng S, Ren Y, Cheng TT et al (2012) Damping of pipeline system using tuned mass dampers. J Vibr Measur Diagn 32(5):823–826
8. Li B, Niu WC, Xu ZY (2016) Eddy current vibration absorber design and experiments. J Northwest Polytech Univ 1:18–24
9. Habib G, Kerschen G (2015) Suppression of limit cycle oscillations using the nonlinear tuned vibration absorber. Proc R Soc A Math Phys Eng Sci 471(2176):20140976
10. Benacchio S, Malher A, Boisson J et al (2016) Design of a magnetic vibration absorber with tunable stiffnesses. Nonlinear Dyn 85(2):893–911
11. Bonsel JH, Fey RHB, Nijmeijer H (2004) Application of a dynamic vibration absorber to a piecewise linear beam system. Nonlinear Dyn 37(3):227–243
12. Sun HL, Zhang PQ, Chen HB et al (2008) Application of dynamic vibration absorbers in structural vibration control under multi-frequency harmonic excitations. Appl Acoust 69(12):1361–1367
13. Webster AC, Vaicaitis R (1992) Application of tuned mass dampers to control vibrations of composite floor systems. Eng J Am Inst Steel Constr 29(3):116–124
14. Yang C, Li D, Cheng L (2011) Dynamic vibration absorbers for vibration control within a frequency band. J Sound Vibr 330(8):1582–1598
15. Kęcik K, Mitura A, Warmiński J (2013) Efficiency analysis of an autoparametric pendulum vibration absorber. Eksploatacja i Niezawodność 15(3):221–224

16. Yuan L (2007) Optimal design of parameters for dynamic vibration absorber in two-degree-freedom systems. J Hunan Univ Technol 2:46–48
17. Li JF, Gong XL, Zhang XZ et al (2005) Study of adaptive tuned vibration absorber and its dynamic properties. J Exp Mech 20(4):507–514
18. Xu ZB, Gong XL, Chen XM et al (2009) Study on mechanical adaptive tuned vibration absorber. China Acad J Electron Publ House 9:1057–1062
19. Hill S, Snyder S, Cazzolato B (2002) An adaptive vibration absorber. In: Acoustics 2002—innovation in acoustics and vibration. Annual conference of the Australian Acoustical Society. Adelaide, Australia
20. Mirsanei R, Hajikhani A, Peykari B et al (2012) Developing a new design for adaptive tuned dynamic vibration absorber (ATDVA) based on smart slider-crank mechanism to control of undesirable vibrations. IJNEM
21. Gong X, Peng C, Xuan S et al (2012) A pendulum-like tuned vibration absorber and its application to a multi-mode system. J Mech Sci Technol 26(11):3411–3422
22. Aguirre G, Gorostiaga M, Porchez T et al (2013) Self-tuning dynamic vibration absorber for machine tool chatter suppression
23. Mikułowski G, Wiszowaty R (2016) Pneumatic adaptive absorber: mathematical modelling with experimental verification. Math Probl Eng 2016(4):1–13
24. Yang K, Zhang YW, Chen LQ et al (2014) Space structure vibration control based on passive nonlinear energy sink. J Dyn Control 3:205–209
25. Wang LH, Gong XL, Deng HX et al (2007) Adaptive tuned vibration absorber based on magnetorheological elastomers and its optimal control. J Exp Mech 22(z1):429–434
26. Kang CJ, Gong XL, Chen XM et al (2012) Control system for an adaptive-active tuned vibration absorber based on magnetorheological elastomers. J Vibr Shock 31(6):27–31
27. Sun SS, Chen Y, Yang J et al (2014) The development of an adaptive tuned magnetorheological elastomer absorber working in squeeze mode. Smart Mater Struct 23(7):075009
28. Herold S, Mayer D (2016) Adaptive piezoelectric absorber for active vibration control. Actuators 5(1):7
29. Rustighi E, Rustighi E, Brennan MJ et al (2003) Design of an adaptive vibration absorber using shape memory alloy 14(1):19–28(10)
30. Gao Q, Fang XB, Zhao YQ et al (2013) Variable mass dynamic vibration absorber and its performance of vibration reduction. J Chang'an Univ (Natural Science Edition) 33(5):109–112
31. Wang BQ (2005) Research on active vibration control technology based on mechanical actuator. Harbin Engineering University
32. Yan C (2007) Research on active vibration isolation technology based on electric actuators. Harbin Engineering University
33. Millitzer J, Ehrt T, Plückhahn A et al (2012) Design, system integration and control concepts of an adaptive active vibration absorber for a convertible. In: ISMA 2012, International Conference on Noise and Vibration Engineering, Conference Proceedings
34. Pagliarulo P, Kuhnen K, May C et al (2004) Tunable magnetostrictive dynamic vibration absorber
35. Konstanzer P, Grunewald M, Jänker P et al (2006) Aircraft interior noise reduction through a Piezo tunable vibration absorber system. Congress of International Council of the Aeronautical Sciences
36. Chen ZL, Ma AL, You XL (2012) Theoretical research of the active-type dynamic vibration absorber. J Xiamen Univ Technol 20(3):67–70
37. Li JQ (2008) Research on active vibration control technology of floating raft system. University of Science and Technology of China
38. Chen B, Li JQ, Shao CX (2008) Experimental study of multi-channel cooperating active vibration control on floating raft. J Exp Mech Anics 23(3):248–254
39. Beltran-Carbajal F, Silva-Navarro G, Abundis-Fong HF (2015) Application of passive/active duffing vibration absorbers in duffing mechanical systems. ICSV22

Chapter 7
Adaptive Frequency Adjustment Control System

Abstract In this chapter, the hardware of controller for ADVA is designed and developed; meanwhile according to the methods of natural frequency adjustment implementation, three control methods are proposed. Through debugging and comparison, the optimal control method and controller design scheme are selected.

7.1 Hardware for Controller

7.1.1 Requirement's Analysis

To make DVA have adaptive tracking control capability, the controller of DVA should be able to decide on whether to adjust the natural frequency and how much to adjust by analyzing the vibration level of the pipeline. At the same time, the controller could output control signal to the actuator for frequency adjustment. In other words, the complete controller should have functions of signal acquisition, signal processing, and control signal output, as is shown in Fig. 7.1. Among them, the signal conditioning circuit needs to complete acquisition signal amplification, filtering as well as power supplying for sensors.

To meet the aforementioned requirements, select the following hardware:

(1) Microprocessor

Due to the requirements in semi-active control application domain and slow change characters of operating conditions, the processor does not need to have a very high dominant frequency and superior computing power; i.e., it is not necessary to use a DSP chip. In addition, due to the interface need for sensors and stepper motors, etc., ordinary microcontroller chips are not applicable for this condition. As a result, microcontroller with ARM core is selected.

(2) A/D conversion chip

Since the vibrations to be controlled are concentrated in the frequency range below 200 Hz and the amplitude change is not significant, the A/D conversion chip

F. Wang et al., *Comprehensive Investigation on Active-Passive Hybrid Isolation and Tunable Dynamic Vibration Absorption*, Springer Tracts in Mechanical Engineering, https://doi.org/10.1007/978-981-13-3056-8_7

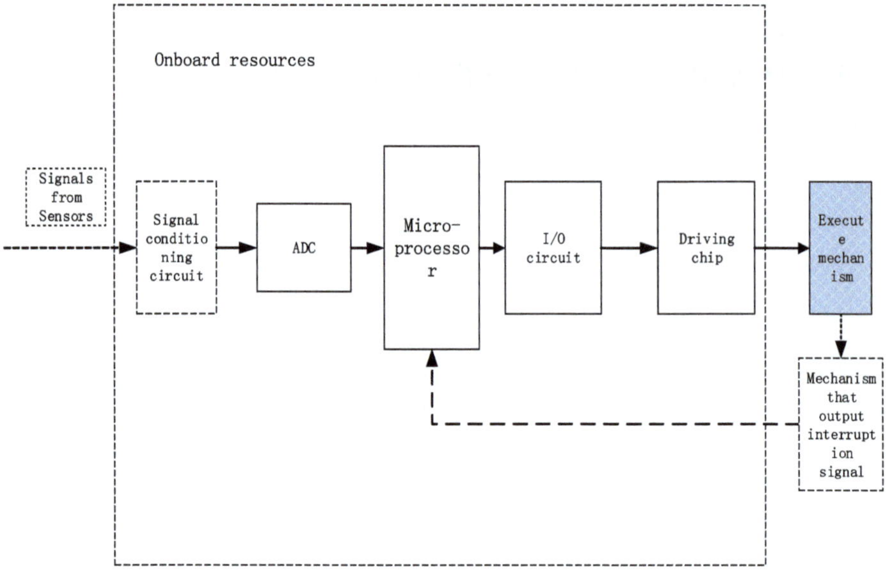

Fig. 7.1 Function modules of controller

is required to have a high conversion accuracy to distinguish the slight amplitude variation while super high conversion speed for the A/D chip is not necessary. What's more, as for three-way vibration of the piping system are concerned, it is necessary to collect vibrations in three directions; i.e., three acquisition channels are required, and one displacement signal acquisition channel for one direction is required; that is, six input channels are required at least.

7.1.2 Chip Selection and Design

After thorough investigation of relevant components on the market, TMC260 chip developed by TRINAMIC, which is a professional motor motion control chip development company in Germany, was selected for step motor driving, as is shown in Fig. 7.2. Due to its small size and low power consumption, this chip could be integrated with microprocessor, ADC conversion chip, etc., on a PCB.

The TMC260 is a dual full bridge driver chip for bipolar stepper motor driving. The internal integration does not require the stall detection function of the sensor. It could be used in position control where the external sensor cannot be installed. This function can also predict the overload condition of the motor and is suitable for applications requiring high reliability. MOSFETs, which were integrated inside the chip, use the unique Low-RDS-ON technology to achieve low power consumption and high efficiency. In addition to strengthen the self-cooling of the motor and driver,

Fig. 7.2 TMC260 stepper motor driver chip

without external cooling equipment, the external environment temperature is high to 1.7 A drive current output. Figure 7.3 shows its schematic.

TMC260's internal integrated DAC function enables micro-step control of current. TMC260 could be controlled through SPI serial interface and STEP/DIR signal input. In addition, the chip also has short circuit, over temperature, under voltage, overload, and other protection functions. The functions are as follows:

(1) Integrated sensor-less stall detection function (Stall Guard) and load measurement function, which are very important for the design of ADVA. It can make the processor know whether the mass reaches the end of the spring rod without the use of external sensors. Because, when the mass reaches the end of the cantilever rod, the stepper motor will stall, i.e., the load will become larger suddenly, and then the stall Guard function of the chip will send an interrupt signal to the processor informing that the mass has reached the end of the cantilever rod. It is no longer to use micro-position switch to achieve this function, thereby reducing

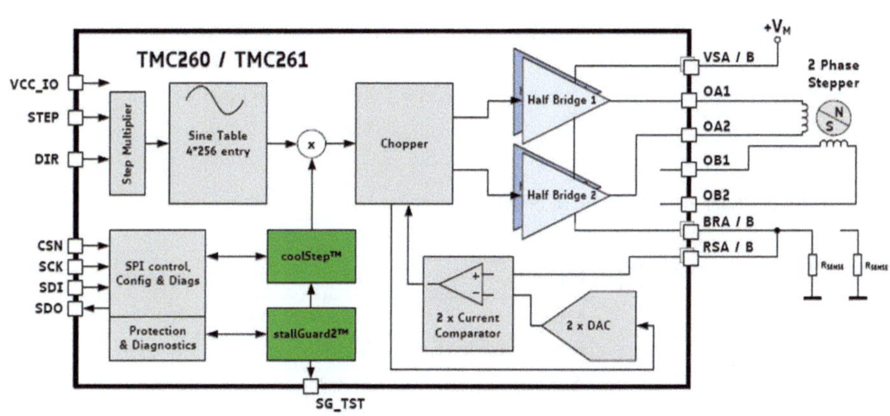

Fig. 7.3 Functional block diagram of TMC260

the design difficulty of the absorber, simplifying the mechanical structure, and making the entire system much more compact.

(2) The cool step function, depending on the load current, can save up to 75% of energy; this is very important for battery-powered systems and can significantly reduce battery capacity and size.

(3) Support 20-bit SPI interface control through simple and convenient SPI serial bus control or STEP/DIR signal control.

(4) Short circuit, over temperature, and overvoltage protection function.

(5) The internal integrated 64-bit DAC can realize 256 times of micro-step subdivision itself, with low-speed smooth control capability; it can realize arbitrary subdivision control through external analog signals.

(6) 7–60 V motor drive voltage, up to 8 A motor drive current, can drive up to 5000 rpm when driving ordinary two-phase stepper motor.

In summary, the overall design scheme is shown in Fig. 7.4. The selected chips and required quantities are listed in Table 7.1.

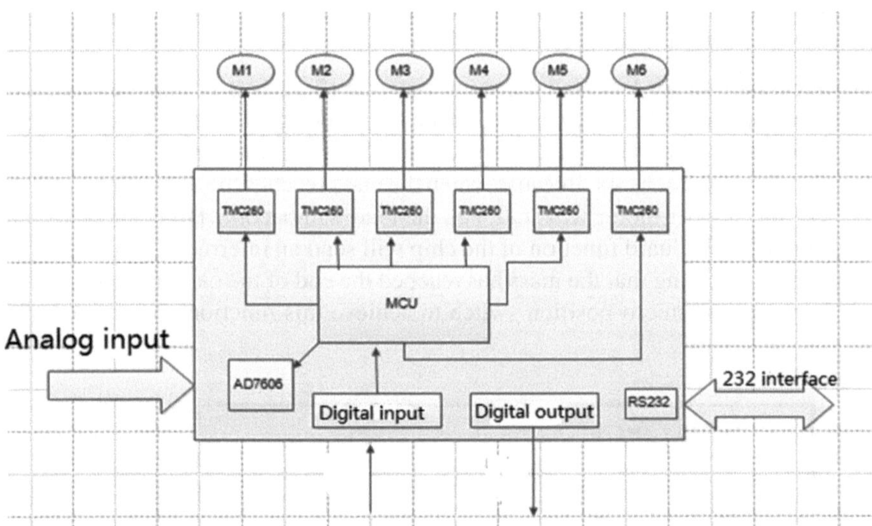

Fig. 7.4 Overall design schematic of controller

Table 7.1 Selected chips' list

Name	Type	Amount
MCU	STM32F407VET6	1
Step motor drive IC	TMC260	6
ADC	AD7606	1
RS232	MAX3232	1
Optocoupler	TLP290	8
Power chip	Tps54331	1

Among them, the microprocessor uses ST's ARM Cortex-M4 core STM32F407 M4 floating-point processor, clocked at 168 MHz, on-chip contains 196K RAM, 1M FLASH; using the Huff pipeline type, internal integration FPU, DSP instructions, TTL serial port, TTL switch input and output control. Using FSMC hardware interface and AD7606 parallel communication, with high-speed and reliable advantages, internal integrated FPU DSP instructions can speed up the implementation of FFT algorithm, through the RS232 can communicate with host computer, and through the optocoupler TTL digital input interface can isolate external input signal.

The AD conversion module is intended to use the Analog Devices AD7606 chip. The maximum sampling voltage range of the chip is [−10, +10 V], the sampling rate is up to 200 KSPS, 16-bit sampling accuracy, each chip has eight independent sampling channels, and each sampling channel could be converted at the same time. The AD sampling scheme is configured and read by the MCU. The AD sampling frequency is configured to 200 KSPS.

The prototype of controller is shown in Fig. 7.5. After debugging, it was found that high-frequency interference may occur between the lines, making the controller unable to work stably, after adjustment of wiring. The PCB is shown in Fig. 7.6. The prototype is shown in Fig. 7.7, which operates stably at both high and low frequencies.

7.2 Principle of Frequency Adjustment

The function of control system is to adjust the position of the slider mass by controlling the rotation direction of the stepper motor and the number of rotation steps so that the natural frequency f of the three directions of the ADVA could match the frequency f_{max} corresponding to the maximum amplitude of the three directions of the pipeline.

Assume that the stepper motor rotates one revolution which could bring the slider forward or backward 0.001 m.

The calculation formula for f is as follows:

Fig. 7.5 Prototype of controller

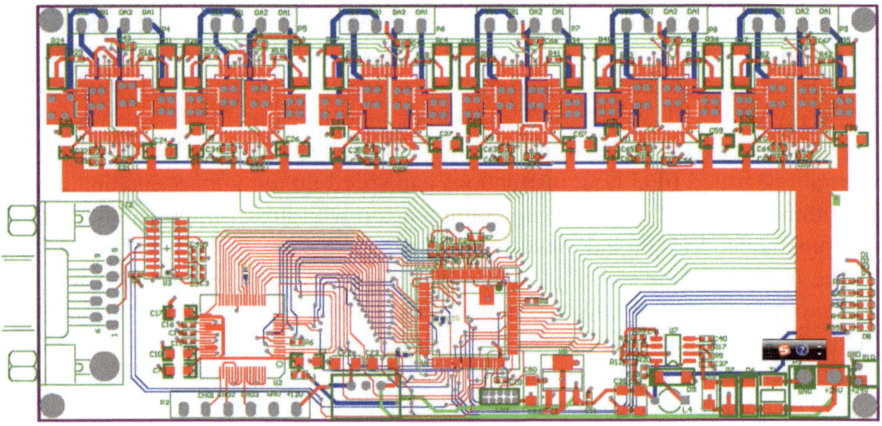

Fig. 7.6 PCB layout of six-stepper motor board schematic

Fig. 7.7 Prototype of six-step motor version controller

$$f = \sqrt{a/l^3} \quad \text{(Simplified formula)} \tag{7.1}$$

where l is the distance from the initial position after the stepper motor has moved.

Assuming in the initial system

$$f = f_0 \tag{7.2}$$

i.e.,

$$f_0 = \sqrt{a/l_0} \tag{7.3}$$

Therefore, the coefficient could be obtained, i.e.,

$$a = f_0^2 l_0 \tag{7.4}$$

So, if after the nth cycle,

$$f\mathrm{max} = f\mathrm{max}\_n \tag{7.5}$$

Currently, the position of the motor is

$$l = l_{n-1} \tag{7.6}$$

According to

$$f\mathrm{max}\_n = f = \sqrt{a/l_n^3}, \tag{7.7}$$

the position where the motor should be is

$$l_n = \sqrt[3]{a/f^2_{\mathrm{max}\_n}} \tag{7.8}$$

Substitute formula (4.42) into

$$l_n = \sqrt[3]{f_0^2 l_0/f^2_{\mathrm{max}\_n}} \tag{7.9}$$

The distance how much the motor should move is

$$\Delta l = l_n - l_{n-1} \tag{7.10}$$

The number of turns that the motor needs to run is

$$\mathrm{count} = \Delta l/0.001 \tag{7.11}$$

Therefore, according to the frequency division number, the number of pulses for driving the stepping motor could be obtained.

In summary, variables that need to be known in advance are l_0, f_0 and the frequency division number, as for the other variables, they all could be deduced from these three variables.

7.3 Adaptive Frequency Adjustment Strategies

According to the manner of adjusting the position of slider, three adaptive frequency adjustment strategies could be developed:

(1) Drive the step motors to adjust the natural frequencies of ADVA to traverse in three directions. Meanwhile, observe vibration changes in the corresponding direction of the pipeline in the process of adjusting natural frequencies of the ADVA from the lowest to the highest, respectively. The frequency corresponding to the minimum vibration level is the frequency that the absorber should have, as is shown in Fig. 7.8.

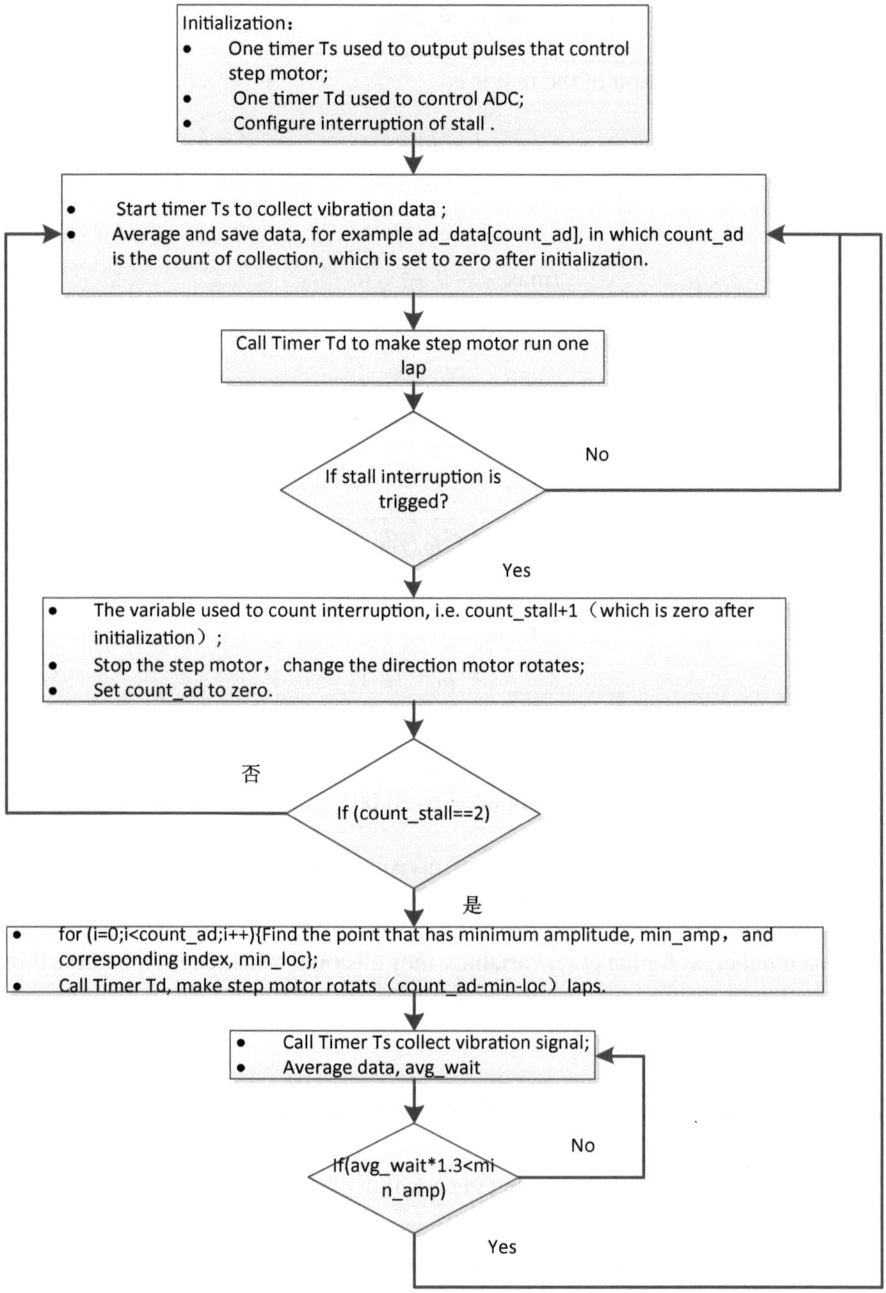

Fig. 7.8 Control flow chart

(2) After the ADVA has been assembled, calibrate and store the relationship between the natural frequency of the vibration absorber in three directions and the position of the slider by frequency sweep test. After being applied to the piping system, the frequency corresponding to the maximum amplitude of the pipeline in three directions could be known by collecting vibration data in three directions of the pipeline and performing FFT calculation. Then, the position of the slider should be located could be acquired through lookup table method. Therefore, drive the stepper motor to adjust the slider to this position, and best vibration absorption performance could be acquired, as is shown in Fig. 7.9.

(3) Firstly, like the procedures in method one, adjust the ADVA to traverse in the three-direction adjustable frequency range one after another to determine the position of the slider when the vibration in the corresponding direction of the pipeline is minimum. The vibration data in this direction are subjected to FFT operation, and the corresponding amplitude is extracted. The frequency and the slider position are stored as a set of data. When the absorber vibration frequency needs to be changed due to changes in the pipe vibration, the table is first checked to see if the frequency is already in the previously saved table, and the frequency position correspondence moves the slider directly to this position; otherwise, it repeats the previous operation. That is, to give the controller the ability of machine learning it is possible to gradually establish the corresponding relationship between the natural frequency of the vibration absorber and the slider position in the second operation of the system and to shorten the time required for the slider adjustment due to the working conditions. After a certain period of time, the relationship between frequency and position is corrected for changes in characteristics caused by creep of the system, maintaining frequency adjustment accuracy and ensuring shock absorption performance, as is shown in Fig. 7.10.

7.3.1 *Transverse*

The specific implementation steps of method one are as follows:

Step 1: Start, mainly for the initialization of the program, and check whether the power is normal or not.

Step 2: Collect the X-direction vibration signal currently and perform the operation of RMS and save results.

Step 3: Drive the stepper motor rotate one revolution to move the slider.

Step 4: Judge whether there is an interruption (indicating that the slider has moved to the end of the cantilever beam, achieved by the position sensor) occurs; if yes, jump to step 5; otherwise, go to step 3.

Step 5: Add 1 to the stall count and change the direction of step motor operation.

Step 6: Determine if the stall count is 2, if yes go to step 2, otherwise, go to step 7.

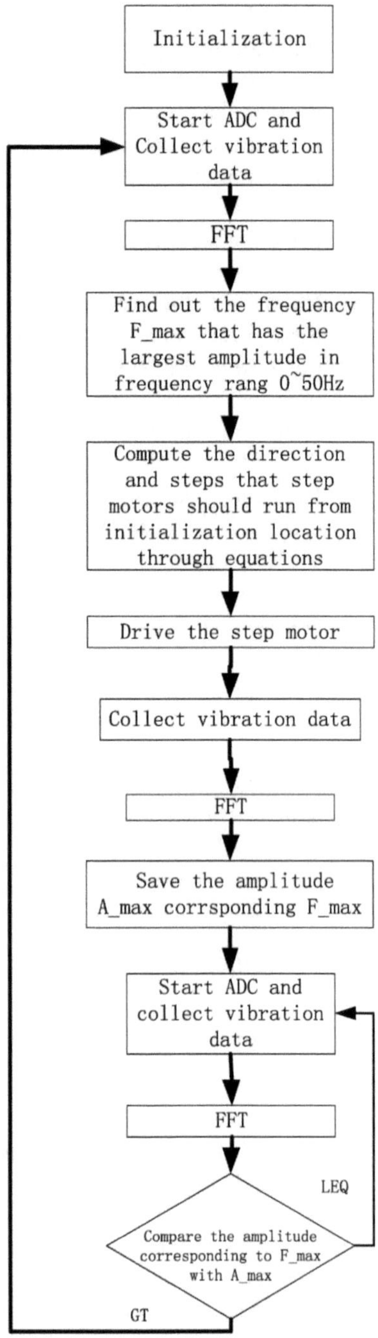

Fig. 7.9 Control flow chart

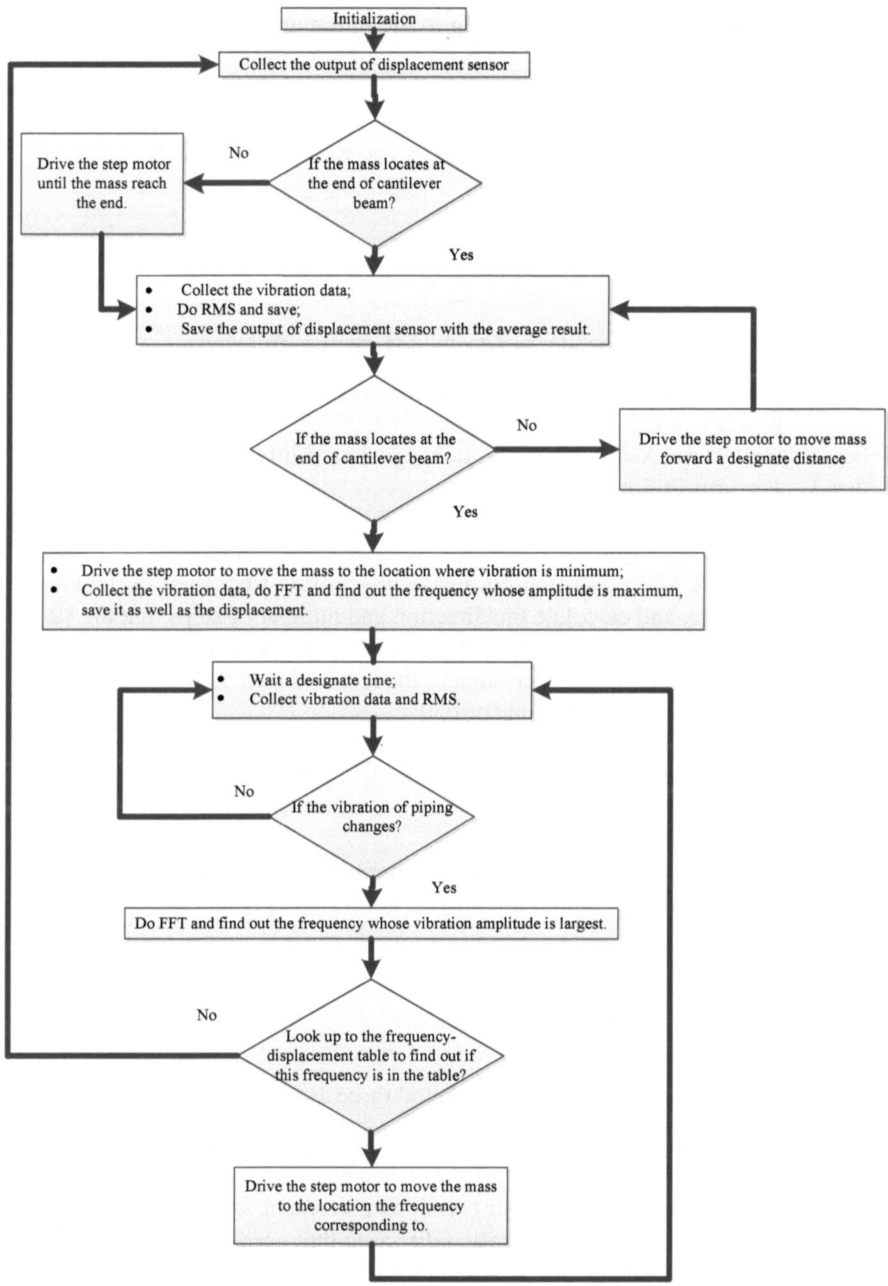

Fig. 7.10 Control flow chart

Step 7: Find the position corresponding to the minimum amplitude and move the slider to this position.
Step 8: Collect the vibration signal currently, perform the RPM operation, and save results.
Step 9: Judge if the amplitude changes, if yes go to step 2; otherwise, go to step 8.

7.3.2 *Lookup Table*

The specific implementation steps of method two are as follows:

Step 1: Start, mainly for the initialization of the program, and check whether the power is normal or not.
Step 2: Collect the X-direction vibration signal currently.
Step 3: Perform FFT operation.
Step 4: Find out the frequency corresponding to the maximum amplitude in this frequency range.
Step 5: Find the slider position corresponding to this frequency by looking up the table, and calculate the direction and number of steps that the stepper motor needs to run to adjust the slider to this position.
Step 6: Drive stepper motor according to the result of step 5.
Step 7: Collect the vibration data currently.
Step 8: Perform FFT operation.
Step 9: Find the maximum amplitude currently and save it.
Step 10: Collect vibration data.
Step 11: Perform FFT operation.
Step 12: Determine whether the maximum amplitude currently is greater than the previously saved amplitude; if yes, go to step 2; otherwise, go to step 10.

7.3.3 *Machine Learning*

The specific implementation steps for method three are as follows:

Step 1: Start, mainly for the initialization of the program, and check whether the power is normal or not.
Step 2: Acquire the signal of X-direction displacement sensor, in which the X-direction refers to the vibration direction that needs to be controlled.
Step 3: Determine whether the X-direction slider is located at the endpoint. If yes, proceed to step 4; otherwise, jump to step 5.
Step 4: Acquire X-direction vibration data in time T, calculate the mean square sum, and save. Save the X-axis displacement meter current data and perform a one-to-one correspondence with the RMS data.
Step 5: Drive the stepper motor until the slider is at the end.

Step 6: Acquire the displacement meter signal to judge whether the slider is located at the end, if yes, go to step 7; otherwise, go to step 8.

Step 7: Drive the X-direction stepping motor to move the slider to the position where the vibration level is minimum; collect the current vibration data and do FFT to find out the frequency corresponding to the maximum vibration and acquire the current signal of the displacement meter. Save in frequency–displacement correspondence table.

Step 8: Drive the X-direction stepper motor to advance the mass move by a specified distance.

Step 9: Wait for a specified time; collect the X-direction vibration data and do the operation of RMS.

Step 10: Determine whether the pipeline vibration conditions changes, if yes, go to step 11, otherwise return to step 9.

Step 11: Perform FFT on the vibration data and find out the frequency corresponding to the maximum amplitude.

Step 12: Compare with the stored frequency–displacement table to see if there is such a frequency. If yes, go to step 13; otherwise, go back to step 2.

For the first design solution, the code for the standing wave detection of the TMC260 stepper motor driver chip consists of two parts: the drive current and the out-of-step threshold, as is shown in Fig. 7.11. The debugging found that due to the existence of irremovable friction existed in the mechanical part, and with the increase of the system operating time, the torque that the stepper motor needs to overcome also changes, which makes it difficult to adjust the drive current and the out-of-step threshold to the proper state, resulting in two situations:

(1) The stepping motor will continuously generate interrupt signals during operation, that is, false alarm.
(2) Even though the slide mass has been moved to the end of the cantilever rod, i.e., it is in a stuck state, and it cannot generate an interrupt signal as required, that is, a missed report.

For the second design scheme, debugging found that due to stepping motor out of step and the existence of indelible clearances in the screw and nut, the position error of the entire system will increase with the increase of the system running time; therefore, in order to ensure the accuracy of frequency adjustment, it was strongly suggested that the position corresponding to the natural frequency should be calibrated periodically which inevitable increases the workload of post-maintenance.

For the third scheme, during the test it was found that due to the introduction of displacement sensors, the accuracy of slide position control, i.e., the accuracy of fre-

```
u8 TMC260Flags = 0;
//负载检测和电流设置寄存器初始化
NO1_StallGuardConfig.FilterEnable = 0;                  /*失步检测滤波0:标准模式, 1:4个fullstep滤波*/
NO1_StallGuardConfig.StallGuardThreshold= 22;           /*带符号失步门限-64~63, 值越大灵敏度越低, 不建议小于-10*/
NO1_StallGuardConfig.CurrentScale =5;                 /*驱动电流0~31*/
```

Fig. 7.11 Adjustment codes of TMC260 standing wave detection

quency tuning completely, depends on the accuracy of displacement sensors, which greatly improves the control accuracy. Meanwhile, in the operation of the algorithm, it could be greatly assisted by machine learning, which substantially reducing the time required for the adaptive adjustment of ADVA under the same conditions.

In summary, the third design scheme can not only guarantee the running accuracy within the entire system operation cycle, but also can greatly reduce the time required for adaptive vibration tuning because of the changes of working conditions. Therefore, the third control strategy is selected as the control strategy for semi-active vibration absorbers.

7.4 Conclusion

A vibration absorber control system was established and three control methods were proposed according to different way to adjust frequency. The optimal control method and controller design scheme were selected through debug and test.

As for functions that control system requires, requirement's analysis, design, hardware selection, and manufacture are operated step by step.

According to different ways of adjusting the slider, three different control methods are proposed. By comparing the time and adjustment accuracy required for frequency adjustment meanwhile considering variable working conditions, a control method with self-learning ability is selected. This method establishes the correspondence relationship between the position of the correcting slider and the natural frequency of the vibration absorber during system operation, which can not only ensure the accuracy of frequency adjustment in the entire system operation cycle, but also greatly reduce the time needed for adaptive frequency adjustment due to the change of operating conditions.

Chapter 8
Experimental Verification for ADVA

Abstract This chapter studies the application of three-way adaptive dynamic vibration absorber in a typical piping system of vessel, including the vibration characteristics analysis of a typical piping system of vessel, the vibration absorption effect test under single-frequency excitation, and the vibration absorption effect under the actual working conditions.

8.1 Introduction

The previous chapter studied the design technology based on the characteristics of pipeline vibration attenuation. The designed pipeline three-direction ADVA is compact and integrated and has advantages of stable performance, high reliability, and wide adjustable frequency range. At the same time, machine learning control strategy based on the adaptive vibration absorption characteristics of pipeline can greatly reduce the operation time required for frequency adjustment while ensuring control accuracy. The vibrating table was used to test the frequency adjustable ability of the designed ADVA in three directions. Results show that the designed ADVA had a wide adjustable frequency range in all three directions of X, Y, and Z.

This chapter will further verify the vibration absorption effect of ADVA in a typical piping system of ship including the analysis of the vibration characteristics of a typical vessel piping system and the single-frequency excitation vibration absorption effect test and vibration absorption effect test under actual working conditions.

8.2 Test Environment

The integrated pipeline test platform consists of a large-scale compartment structure, a sea and oil-cooling system (including machinery and piping), and a pressure tank, as is shown in Fig. 8.1.

F. Wang et al., *Comprehensive Investigation on Active-Passive Hybrid Isolation and Tunable Dynamic Vibration Absorption*, Springer Tracts in Mechanical Engineering, https://doi.org/10.1007/978-981-13-3056-8_8

Fig. 8.1 Comprehensive test platform of piping system vibration and noise

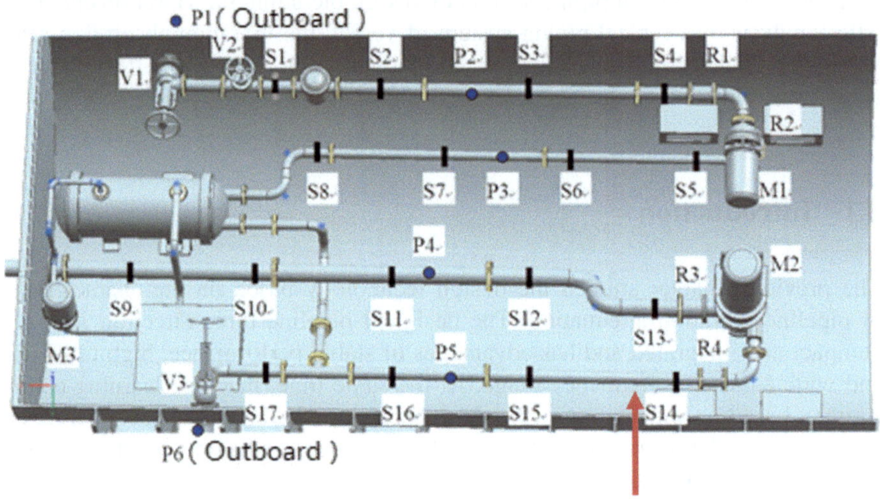

Fig. 8.2 Mechanical system equipment and corresponding component numbering

Number the equipment in the piping system, as is shown in Fig. 8.2, and select part of the locations to perform the vibration test of the piping system. The position indicated by the red arrow in the figure is the installation location of ADVA.

It could be found that the 50 Hz peak of the pump vibration blade is prominent. Therefore, to control the vibration transmitted from the pipeline to the housing, attention must be paid to low frequency vibration. That is, it is necessary to take targeted measures on the frequencies at which low-frequency line spectrum stands out.

Fig. 8.3 Way to excite piping system

8.3 Test Under Single-Frequency Excitation

To fully verify the vibration absorption performance of ADVA, the vibration absorption performance of ADVA under single-frequency exciting is first conducted.

8.3.1 Test Method

The way to install the shaker is as is shown in Fig. 8.3. The single-frequency sinusoidal excitation signal used to drive the exciter ranges from 30 to 50 Hz. Measure and record the vibration of X-, Y-, and Z-direction of the piping system as well as the vibration of hull.

The sensors used to measure the vibration of pipe and hull are installed as is shown in the red circles in Figs. 8.4 and 8.5. The test is divided into three procedures:

(1) Measure the vibration of pipe and hull with no vibration absorber installed.
(2) Install ADVA, but lock the spring rod to use ADVA as a mass block, as is shown in Fig. 8.6, and measure the vibration of pipe and hull.
(3) Remove the locking device and turn on the vibration absorber, as is shown in Fig. 8.7, and measure the vibration of pipe and hull.

Fig. 8.4 Location of sensor used to measure the three-direction vibration of piping system

8.3.2 Test Result Analysis

Some of the test results are shown in Figs. 8.8, 8.9, 8.10, and 8.11. It could be found that:

(1) When the ADVA's spring rod was locked, i.e., ADVA is used as an extra mass to piping system, vibration of some frequencies in *X* and *Y* directions was slightly reduced, while the majority vibration was magnified, especially at 350 and 450 Hz in Figs. 8.8 and 8.9, increased by 25 and 15 dB, respectively. As for the *Z*-direction of the piping system, vibration reduced about 10 dB for the frequency band below 50 Hz. However, vibration was magnified by 5–15 dB when frequencies were higher than 450 Hz.

(2) Remove the lock mechanisms, and turn on ADVA; vibration reduction was not only found at the disturbance frequency, but also could be observed in a wide frequency band like that in the *Z*-direction at procedure one. For example, for the vibration in the *Z*-direction of the pipeline, an average of 20 dB vibration attenuation effect was achieved in the frequency band below 50 Hz, as is shown in Fig. 8.10. This is due to the damping properties of ADVA (more details can refer to the sweeping test in Chap. 6; a wide resonance peak was observed during the test, as is shown in Figs. 6.21, 6.22, etc.).

(3) It could be seen from Fig. 8.11 that ADVA has a weak control effect at the disturbance frequency, about 1 dB, and there is almost no control effect at other

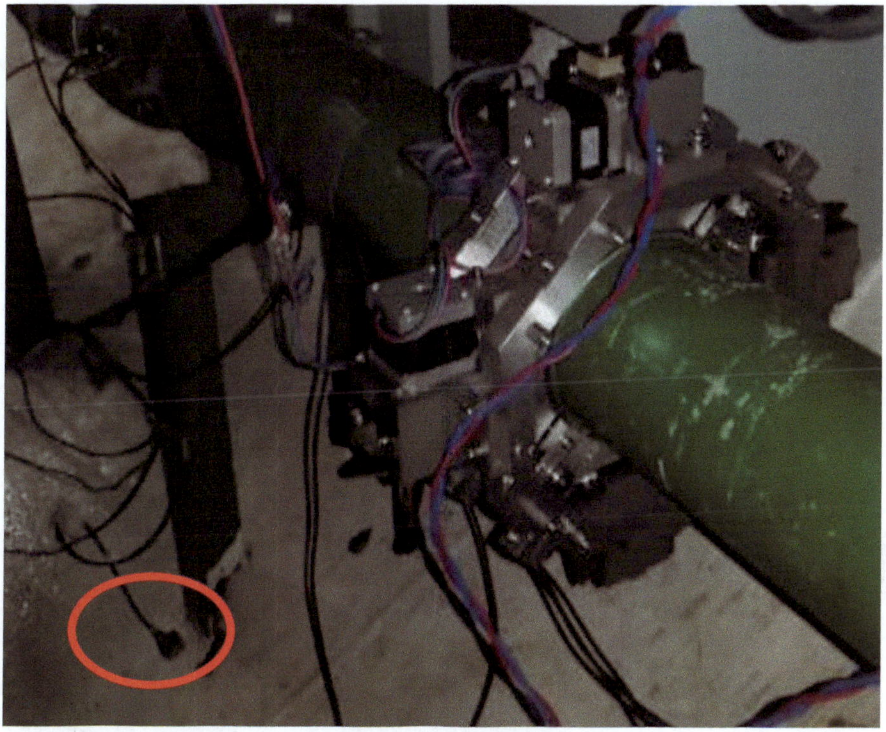

Fig. 8.5 Location of sensor used to measure the vibration of hull

frequencies. One possible reason for this is the shaker transfer vibration directly to the hull through supporting mechanism.

8.4 Test Under Actual Working Conditions

8.4.1 Test Method

Remove the shaker and measure the three-direction vibration of piping system and hull under the condition that bump runs ranges from 1800 to 3000 rpm with the help of a frequency modulator, as is shown in Fig. 8.12. The test also has three procedures:

(1) Measure the vibration of pipe and hull with no ADVA installed.
(2) Install the ADVA, but lock the spring rod to use ADVA as a mass block, as is shown in Fig. 8.13, and measure the vibration of pipe and hull.
(3) Remove the locking device and turn on the vibration absorber, as is shown in Fig. 8.7, and measure the vibration of pipe and hull.

Fig. 8.6 One of the mechanisms used to lock the spring rods of ADVA

8.4.2 Test Result Analysis

Vibration changes happened in the X-direction at different speeds are shown in Figs. 8.13 and 8.14, respectively. It could be found that:

(1) When ADVA was used as an extra mass to piping system, vibration at the frequencies (1800 rpm-30 Hz, 2520 rpm-42 Hz) indicated by pump's rotating speed reduced 10 and 5 dB, respectively. But, as for vibrations at other frequencies, they were magnified basically.
(2) After the adaptive vibration absorption function is activated, ADVA not only attenuated vibration corresponding to the excitation frequency at an average control effect of 7 dB, but also reduced vibration at some other frequencies.
(3) One thing should be noted is that after the adaptive vibration absorption function is enabled, even though ADVA has no attenuation effect at some frequencies, it would not amplify them either.

Figures 8.15 and 8.16 show the changes happened in hull's vibration when the pump is rotating at speed of 1800 and 2520 rpm, respectively. It could be seen from the figure that:

Fig. 8.7 ADVA without lock mechanisms

(1) When the ADVA did not work, the ADVA had no influence on the hull at most frequencies; even if a few frequencies had vibration reduction effects, such as at frequency 250 Hz in Figs. 8.15 and 8.16, 8 and 13 dB vibration reduction were achieved, respectively. Meantime, vibration amplification phenomena could be observed at some frequencies, such as frequency of 90 and 80 Hz in Figs. 8.15 and 8.16, and the amplification is about 2 dB.
(2) After the adaptive vibration absorption function is enabled, vibration at some frequencies were attenuated. An interesting thing is that this vibration reduction did not happen at disturbance frequencies, such as comparative vibration reduction happened at 240 and 250 Hz in Figs. 8.15 and 8.16, reduced about 7 and 12 dB, respectively.

Figures 8.17 and 8.18 show the influence of ADVA on the vibration in the Y-direction of the piping system when the pump is rotating at speed 2640 and 2760 rpm, respectively. We can find that:

(1) Compared with those happened in the X-direction of the piping system, vibration reduction could hardly be observed in the Y-direction at disturbance frequency (2640 rpm-44 Hz, 2760 rpm-46 Hz).

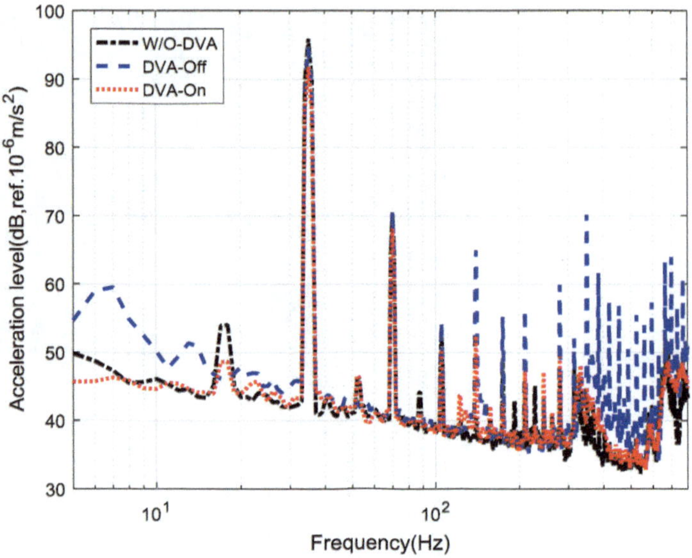

Fig. 8.8 Comparison of vibration levels in the X-direction before and after vibration absorption (disturbance frequency was 35 Hz)

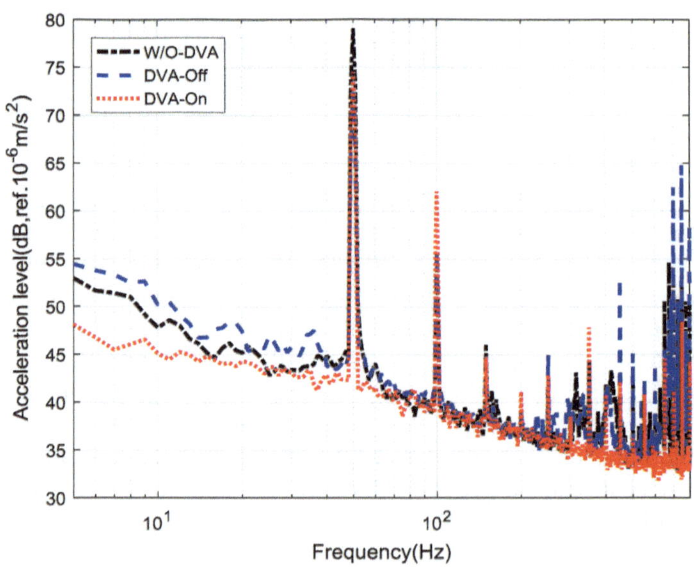

Fig. 8.9 Comparison of vibration levels in the Y-direction before and after vibration absorption (disturbance frequency was 50 Hz)

(2) In the Y-direction, ADVA used as extra mass to pipe and also reduced and increased the vibration at other frequencies, although the majority were mag-

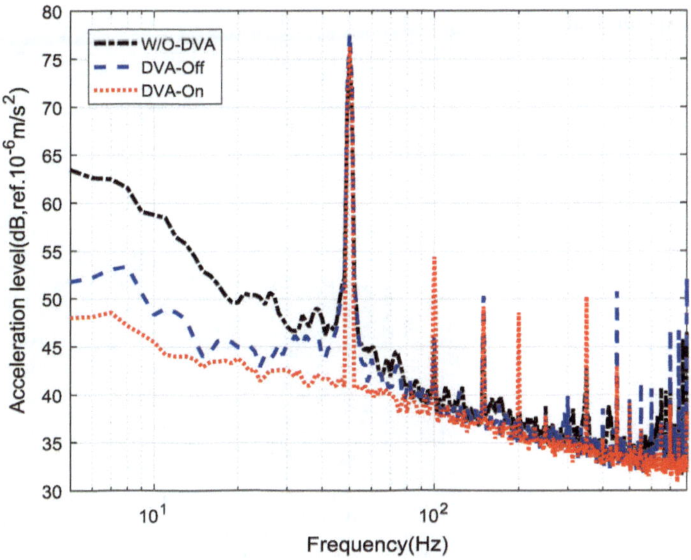

Fig. 8.10 Comparison of vibration levels in the Z-direction before and after vibration absorption (disturbance frequency was 50 Hz)

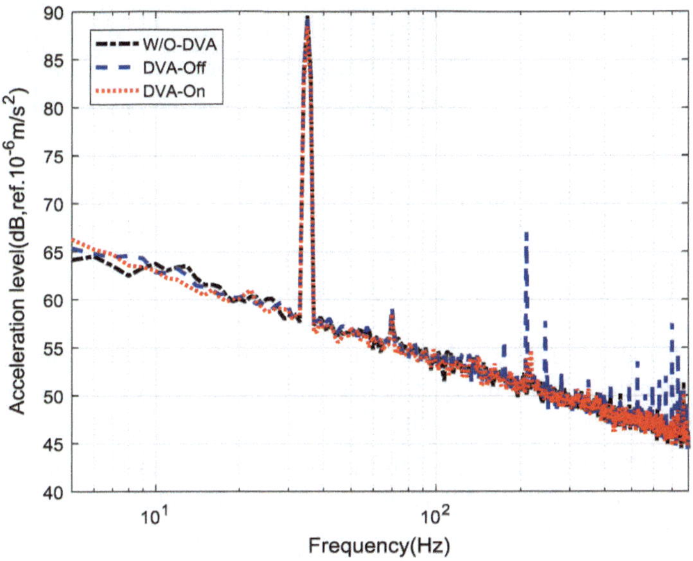

Fig. 8.11 Comparison of the vibration of hull before and after vibration absorption

nified. For example, control effects of 3 and 8 dB for 45 Hz in Fig. 8.17 and 170 Hz in Fig. 8.18 could be observed, respectively, while vibration magnifies to 8 dB for 150 Hz in Fig. 8.17 and 250 Hz in Fig. 8.18, respectively.

Fig. 8.12 Experimental layout under actual conditions

(3) After the adaptive vibration absorption function is activated, the vibration absorber has a certain attenuation effect on the vibration corresponding to the excitation frequency, which is about 3 dB. Obviously, it is not as good as the vibration absorption effect in the X-direction (7 dB).
(4) After the adaptive vibration-absorbing function is enabled, the vibration absorber also attenuates vibration at some other frequencies in the Y-direction. For example, at 20 Hz in Fig. 8.17, the control effect is 5 dB. And even if there is no attenuation, it will not amplify the vibration, just like that in the X-direction.

Figures 8.19 and 8.20 show the influence of ADVA made on hull when the pump is rotating at speed of 2640 and 2760 rpm, respectively. It could be found that vibration at most frequencies of the hull was hardly changed, even though significant vibration reduction could be observed at 70 and 250 Hz in Figs. 8.19 and 8.20, and control effects were 2 and 13 dB, respectively.

Figures 8.21 and 8.22 show the influence that ADVA made on the Z-direction when the pump is rotating at speed of 2640 and 3000 rpm. It could be found that:

(1) When ADVA was used as an extra mass to the piping system, it has little affection on the vibrations corresponding to disturbance frequencies.

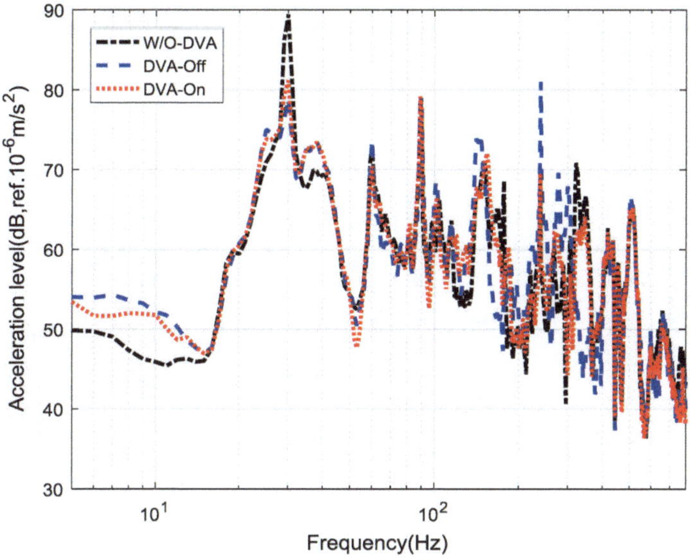

Fig. 8.13 Comparison of vibration levels in the X-direction before and after vibration absorption (the pump rotating speed is 1800 rpm)

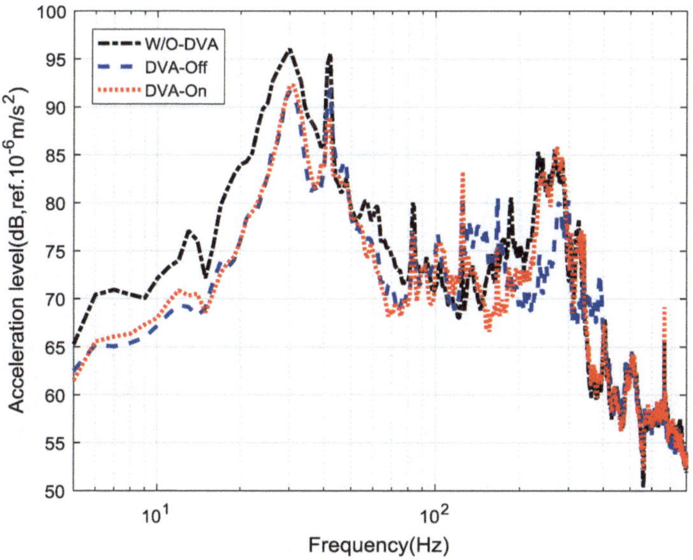

Fig. 8.14 Comparison of vibration levels in the X-direction before and after vibration absorption (pump rotates at 2520 rpm)

(2) ADVA used as mass had better vibration control effect at certain frequencies than the adaptive absorption function enabled, such as frequencies 50 and 100 Hz in

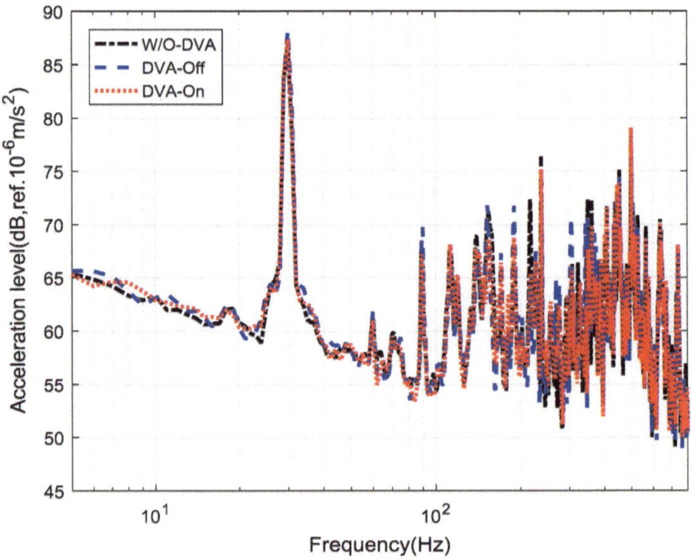

Fig. 8.15 Comparison of hull vibration levels before and after vibration absorption (pump rotates at speed 1800 rpm)

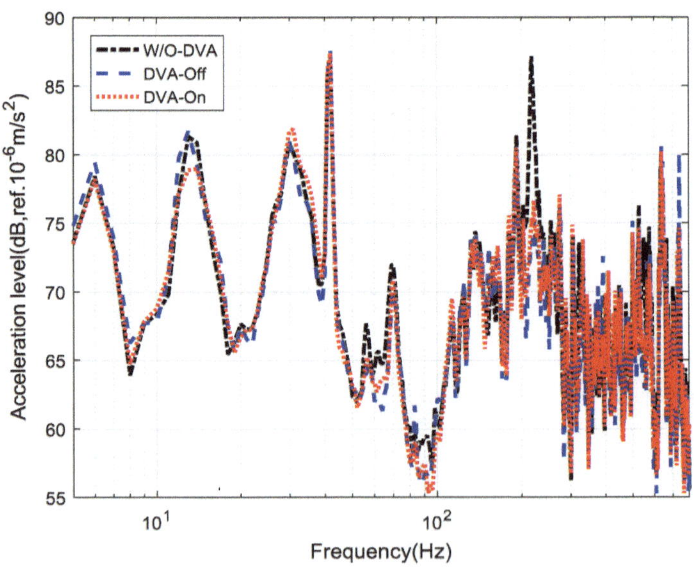

Fig. 8.16 Comparison of hull vibration levels before and after vibration absorption (pump rotates at speed 2520 rpm)

Fig. 8.21, and the vibration reduction effect were 5 and 2 dB better, respectively; 170 and 220 Hz in Fig. 8.22 were 2 and 8 dB better.

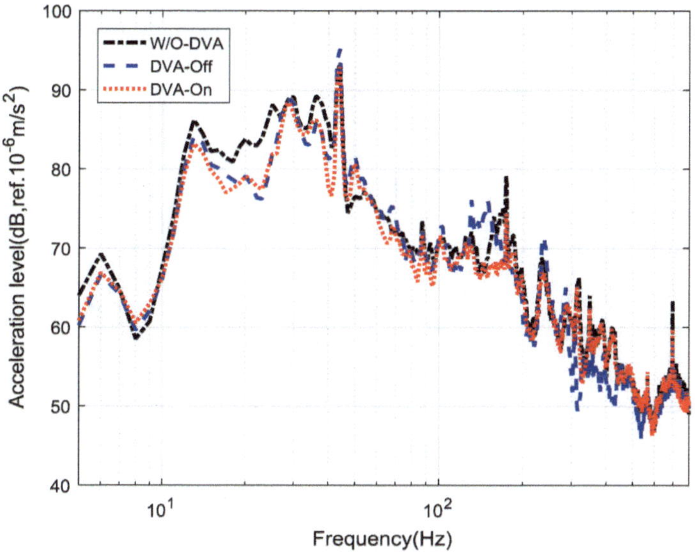

Fig. 8.17 Comparison of vibration levels in the *Y*-direction before and after vibration absorption (pump rotates at 2640 rpm)

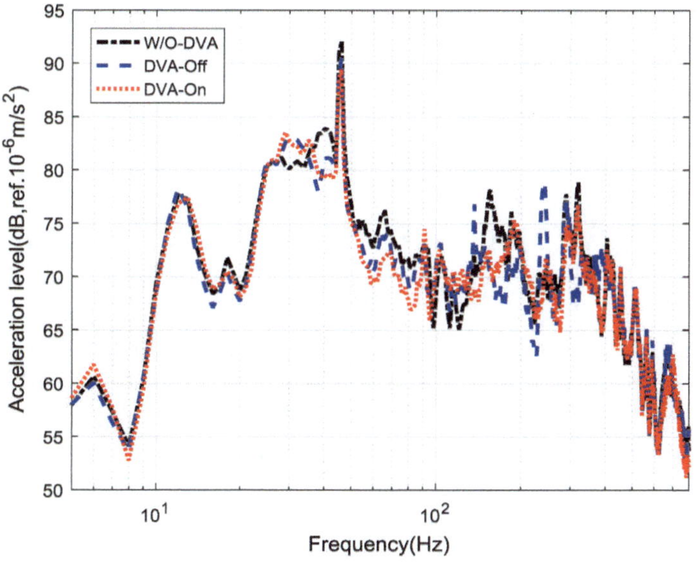

Fig. 8.18 Comparison of vibration levels in the *Y*-direction before and after vibration absorption (pump rotates at 2760 rpm)

(3) After the adaptive vibration absorption function is activated, ADVA had an average attenuation effect of 6 dB on the disturbance frequency.

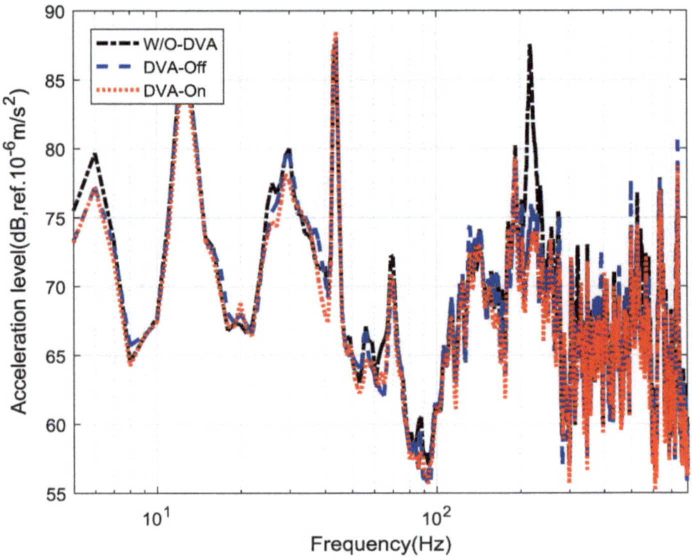

Fig. 8.19 Comparison of hull vibration levels before and after vibration absorption (pump rotates at 2640 rpm)

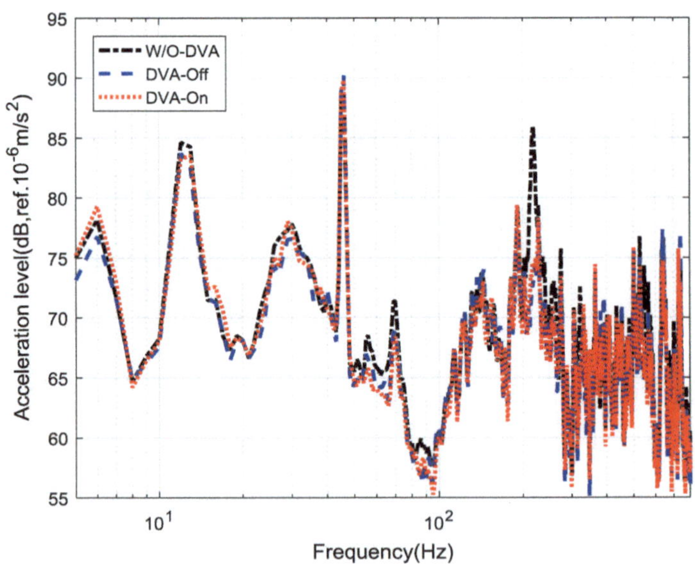

Fig. 8.20 Comparison of hull vibration levels before and after vibration absorption (pump rotates at 2760 rpm)

(4) After the adaptive vibration absorption function is activated, vibration at frequencies above 100 Hz was amplified.

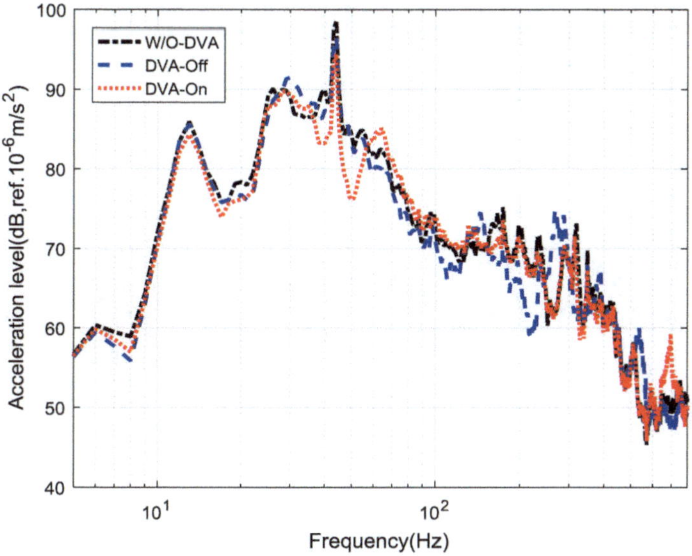

Fig. 8.21 Comparison of vibration levels in the *Z*-direction before and after vibration absorption (pump rotates at 2640 rpm)

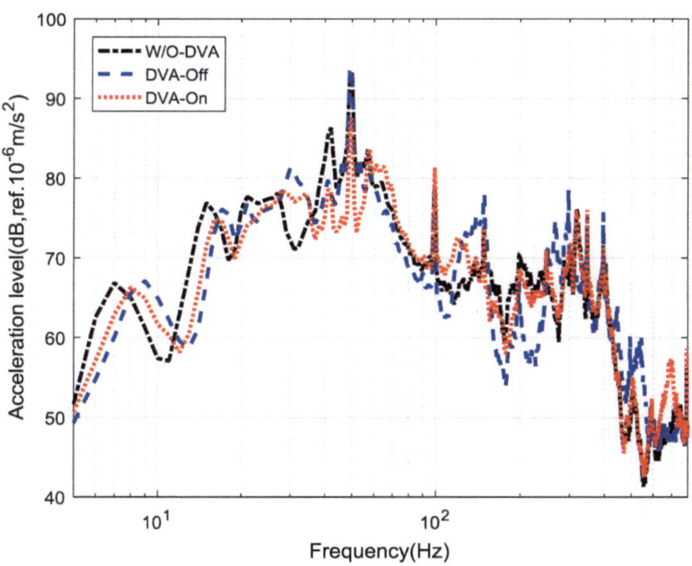

Fig. 8.22 Comparison of vibration levels in the *Z*-direction before and after vibration absorption (pump rotates at 3000 rpm)

Figure 8.23 shows the effect that ADVA made on the vibration of hull when the pump run at a speed of 3000 rpm. It could be found that ADVA could reduce

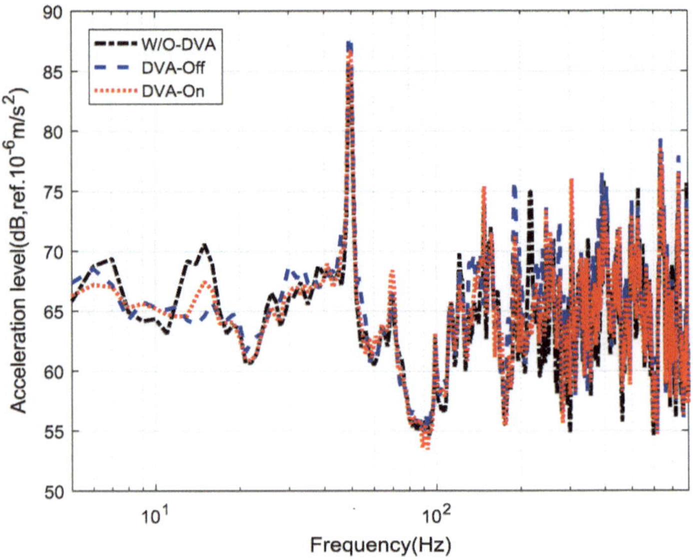

Fig. 8.23 Comparison of hull vibration levels before and after vibration (pump rotates at a speed of 3000 rpm)

the vibration in a frequency band below disturbance frequency. For example, for frequencies below 5 Hz and frequencies from 12 to 17 Hz, an average vibration reduction of 3 dB was achieved. In addition, 3 and 10 dB control effects at 140 and 220 Hz could be observed, respectively. The vibration absorbers that do not start the adaptive vibration absorption function also had a certain attenuation effect on certain frequencies of the vibration of hull. For example, a damping effect about 1 dB for the frequency band of 2–5 Hz, a damping effect about 5 dB for the frequency band from 12 to 17 Hz, and 1 and 7 dB control effects for 140 and 220 Hz could be observed, respectively.

8.5 Conclusion

In this chapter, the designed three-degree-of-freedom ADVA is applied to a typical piping system of naval vessel. The vibrations of piping system and hull are measured with and without ADVA. Experiments of single-frequency excitation and actual work condition verified the vibration absorption effect of ADVA. Research shows:

(1) When the vibration-absorbing function is disabled, i.e., ADVA is used as a additional mass block, the vibration-absorbing device attenuates vibration in three directions of the pipeline and the vibration of hull at certain frequencies, which means increasing the weight of pipeline does not necessarily reduce the

vibration of the pipeline and DVA without self-adaptive capability may amplify vibration if operating conditions change.

(2) When the vibration absorption function is enabled, ADVA can not only achieve adaptive tracking control of the piping system vibration, but also can suppress the vibration of the piping system in a certain frequency band. The average vibration absorption effect for the maximum line spectrum of the piping system is 6 dB. For other line spectral, the vibration reduction capability is also unequal. The maximum control effect is 9 dB happened at 250 Hz line spectrum.

(3) The ADVA designed for reducing the three-direction vibration of piping system is also effective in reducing the vibration of hull, specifically, among which the control effect of the line spectrum of about 250 Hz is more significant, and the control effect is 15 dB.

MIX
Papier aus verantwortungsvollen Quellen
Paper from responsible sources
FSC® C105338

If you have any concerns about our products,
you can contact us on
ProductSafety@springernature.com

In case Publisher is established outside the EU,
the EU authorized representative is:
Springer Nature Customer Service Center GmbH
Europaplatz 3, 69115 Heidelberg, Germany

Printed by Libri Plureos GmbH
in Hamburg, Germany